Nothing in This Bo
But It's Exactly Hoₙ

"Bob Frissell has done it again. His groundbreaking book 30 years ago opened our minds to an amazing explosion of spiritual knowledge. Frissell presents a breathtaking expansion of our understanding and experience of reality with his powerful Five-Step Breath Alchemy healing technique. No better guide exists to the deeper nature of reality."

KENNETH PORTER, M.D., SPIRITUAL PSYCHIATRIST,
PSYCHOANALYST, AND AUTHOR OF *APOLLO'S LYRE*

"Essential reading for anyone drawn to the grand mysteries of life. A complete and necessary guide that delves into humanity's extraterrestrial origins, Earth's true history, different dimensions and levels of consciousness, how to deeply and profoundly heal yourself and your relationships, and more. A must-read for truth seekers and those aspiring to understand the universe and themselves on a deeper level."

KRIS ASHLEY, AUTHOR OF
CHANGE YOUR MIND TO CHANGE YOUR REALITY

"An eye-opener that further unravels mysteries and cover-ups. Bob's divine insight has uncovered hidden truths that raise our consciousness and have the power to change our lives and the world as we now know it. Bob has definitely tuned into the Creator Frequency! If you are searching for truth, I highly recommend this book."

DIANNE ROBBINS, AUTHOR OF
MESSAGES FROM THE HOLLOW EARTH

"If you're looking for a fast-paced, eye-opening, big-pictured, wild-ride of a book, here you go! Bob Frissell sweeps aside the veils of conventional reality to explore what's happening behind the scenes—from extraterrestrial visitations and multidimensional events to walk-ins and ascended masters relating the 'true history of the world'—emphasizing the desperate need for global change and personal transformation."

DAWN BRUNKE, AUTHOR OF
AWAKENING THE ANCIENT POWER OF SNAKE

"I loved Bob Frissell's book *Nothing in This Book Is True, But It's Exactly How Things Are* when it first came out many years ago. The fact that this book has been translated into 25 languages and has sold over 500,000 copies means that Bob is on to something very important. It will make you think, maybe twice, about many things. It is filled with crazy wisdom! It's a genuine gift to all spiritual seekers."

DAN BRULÉ, AUTHOR OF *JUST BREATHE*

"If ever there was an aptly named book, this is it. In a highly readable and sincere manner, *Nothing in This Book Is True, But It's Exactly How Things Are* proceeds to thread together every New Age belief and conspiracy theory into a grand unified field theory of kookiness. They're all here: gray aliens, ascended masters, free energy, angels, cattle mutilations, crop circles, Breath Alchemy, Earth changes, the Great Pyramid, Sirius, and secret colonies on Mars. And yet, despite the sheer unbelievability of half the book, the author's goodwill and spiritual intentions are so infectious that the book ends up being a heartwarming experience."

JAY KINNEY, *WIRED* MAGAZINE

"I highly recommend Bob Frissell's time-tested, monumental book to anyone interested in novel, creative ideas and insights regarding the significance of UFOs in human history. Frissell's book is a concise yet comprehensive study and review of Man's history, his views of Creation, the Creator, and the creature himself, Man. This book is about the expansion of human consciousness and a new understanding that integrates ancient knowledge, myths, and occult traditions with quantum theory and metaphysics to aid the reader in making the crossing from the Piscean Age into an Aquarian Awareness."

ROBERT D. MORNINGSTAR,
EDITOR AND PUBLISHER OF UFO DIGEST

"Bob reminds us that through understanding our origins and going through the transformations necessary to embrace the future, we're somehow going to get past all this dark night and evolve as a people and a planet. I appreciate his lifetime of work with Breath Alchemy as one of the main healing modalities that will help us get there."

DENIS OUELLETTE, AUTHOR OF
HEAL YOURSELF WITH BREATH, LIGHT, SOUND & WATER

"Bob Frissell's book takes readers on an extraordinary journey through the realm of ancient knowledge and the unfolding evolution of human consciousness. Frissell breaks all the dogmas and stereotypes of ideas about our past imposed on us by official science. It is a guide to finding inner peace and creating a brighter future for humanity—a new reality filled with love, unity, and limitless possibilities."

ALEXANDER MILOVANOV, AUTHOR OF
THE MUSIC OF THE DIVINE SPHERES

"The veils have been lifted after reading *Nothing in This Book Is True, But It's Exactly How Things Are*. It's a very enjoyable read which transitions the reader into a quest of self-discovery. It becomes an incredible voyage, undeniably riveting and so intriguing that I couldn't stop reading. I was captivated in such a way that I had never been before! I found myself already upon the index and still wanting more of this newly acquired knowledge on how the universe operates and is operating right now. This book acted as the vital nutrients essential for my growth during this collective spiritual harvest season. This book has changed my life forever."

WILLIAM DAVID, AUTHOR OF
A CHILD'S ADVENTURE

"As you can see in this book, Bob loves to tell the truth. As a result, he has become an internationally famous teacher—and he deserves it."

LEONARD ORR, FOUNDER OF REBIRTHING
AND AUTHOR OF *BREAKING THE DEATH HABIT*

"For us, as lifelong seekers of truth, this book was soul food as our hearts acknowledged the profound truths therein. We loved the way Bob took an enormous amount of information and condensed it into one 'user-friendly' format. A very timely and significant contribution to help us all understand and deal with the personal and global challenges being faced today."

LINDA AND BRENDA MCCOY, COAUTHORS OF
THE LIVING CODE

"This is an accessible guidebook to consciousness expansion and evolutionary progress which, in effect, restates age-old knowledge and in so doing reminds us of our heritage as spiritual beings. . . . Frissell tells his account primarily through the experience of one Drunvalo Melchizedek, a 'walk-in' being who claims to be from the thirteenth dimension by way of several vibration-stepping-down incarnations along the way—on this occasion via a mutual spiritual pact with a male human in 1972. . . . Part of Drunvalo's mission is to remind us that we can . . . assemble within ourselves the universal pattern of creation—our own merkabah vehicles—which will facilitate the evolution of our consciousness as well as that of planet Earth."

NEXUS: THE ALTERNATE NEWS MAGAZINE

"Bob's book reaches back hundreds of thousands, millions of years into the past and sketches out how consciousness has been using third-dimensional reality as a playground and a test lab for life that extends into the lower realms and higher realms for eons! It may well be impossible to explain and understand the vastness of the All That Is, but having a sense of the awe and wonder can make the unfolding experience all the more alive. Eventually we will all awaken. But Bob Frissell's updated metaphysical-spiritual classic *Nothing in This Book Is True, But It's Exactly How Things Are* is like an endless fractal pattern that can open up your awareness to extraordinary, far-out possibilities."

K. SCOTT TEETERS, FORMER HOST
AND PRODUCER OF FAR OUT RADIO

NOTHING IN THIS BOOK IS TRUE, BUT IT'S EXACTLY HOW THINGS ARE

THE SECRET HISTORY OF THE EARTH, MERKABA ACTIVATION, AND BREATH ALCHEMY

A Sacred Planet Book

Bob Frissell

Bear & Company
Rochester, Vermont

Bear & Company
One Park Street
Rochester, Vermont 05767
www.BearandCompanyBooks.com

Bear & Company is a division of Inner Traditions International

Sacred Planet Books are curated by Richard Grossinger, Inner Traditions editorial board member and cofounder and former publisher of North Atlantic Books. The Sacred Planet collection, published under the umbrella of the Inner Traditions family of imprints, includes works on the themes of consciousness, cosmology, alternative medicine, dreams, climate, permaculture, alchemy, shamanic studies, oracles, astrology, crystals, hyperobjects, locutions, and subtle bodies.

Originally published in 1994 by North Atlantic Books under the same title
2nd edition published in 2002 by North Atlantic Books
15th anniversary 3rd edition published in 2009 by North Atlantic Books
Revised and expanded 25th anniversary edition published in 2019 by North Atlantic Books
30th anniversary edition published in 2024 by Bear & Company

Note to the reader: This book is intended as an informational guide. The remedies, approaches, and techniques described herin are meant to supplement, and not be a substitute for, professional medical care or treatment. They should not be used to treat a serious ailment without prior consultation with a qualified health care professional.

Cataloging-in-Publication Data for this title is available from the Library of Congress
ISBN 978-1-59143-518-1 (print)
ISBN 978-1-59143-519-8 (ebook)

Printed and bound in China by Reliance Printing Co., Ltd.

10 9 8 7 6 5 4 3 2 1

Text design and layout by Kenleigh Manseau
This book was typeset in Garamond Premier Pro with Mrs Eaves and Trade Gothic LT Std used as display typefaces
Artwork facing chapters 2, 3, 5, 6, 8, 12 ,13, 14, 15, 16, and 28 is by Spain Rodriguez.
Artwork facing chapters 4, 11, 17, 19, 20, 21, 22, 23, and 26 was created by Jess Kellan using the AI program Midjourney.

Scan the QR code and save 25% at InnerTraditions.com. Browse over 2,000 titles on spirituality, the occult, ancient mysteries, new science, holistic health, and natural medicine.

Contents

Your Word Is the Only Thing That Will Get You Through

Richard Grossinger

In October 1992, I reached a point I could not see my way past or through. My frames of reference—personal, relational, spiritual, cosmic—all seemed either inauthentic or futile. A new issue of *Common Ground*, the Northern California resource guide, lay on the floor in a stack of recycling papers. Without thinking that I was seeking anything concrete I picked it up and opened it almost like a game of divination. At the top of a random page was "East Bay Rebirthing Center," followed by the name Bob Frissell.

I had known about rebirthing for a long time. I had done many ritual and psychospiritual processes similar to rebirthing, and I wasn't particularly attracted to it. Yet at that moment I chose to pick up the phone and call Bob Frissell.

He lived some twenty blocks from me, just over the Oakland line from Berkeley. On the phone he was noncommittal, merely scheduling an appointment. I took notes on how to get there.

I brought my dark cloud with me.

Bob said, almost immediately, "The gap between the way in which you have evolved and the way in which you haven't is getting greater and greater with the years. It is now almost large enough to kill you."

We began with his demonstrating how to breathe. He took breaths at varying speeds and in different rhythms. They all had one component; they were deep, true breaths. They were meant to be pranic winds, blowing through not just the lungs and vessel of his cellular body but his energy bodies as well. His breaths had nothing to do with hyperventilating as an altered state of consciousness. They had to do cleaning out the body, like sweeping the dust and cobwebs from a house that hadn't been cleaned in forty years. They had to do with breathing as letting go, as a trusting of one's higher self and the universe itself.

Bob wanted me to imitate his breathing. He wasn't willing to begin our work until he was sure I could do at least a passable facsimile of it.

It was difficult at first. Some part of me, even in my bleak mood, even with hope of release and revelation, did not want to breathe this hard.

My sessions with Bob consisted solely of this breathing. I felt like a person lighting his own pilot and then summoning a deep fire from it. When I got a blaze going, I sustained its roar as long as I could, traveling on its transformation of oxygen (and, I hoped, prana) into other realms and constellations. It was a delicate, porous, fluttering carpet of energy. It was usually gentle. It always made me so ravenously hungry that, either at the end of the session or during bathroom breaks, I had to coax a few almonds out of Bob's kitchen jar or a slice of bread from the whole wheat loaf sitting beside the knife next to the sink.

The breath journey always let me down gently like a parachute landing from a dream to another dream.

Bob was my guide. He said little. Mostly he might remind me to keep breathing—that is, to warn me when my breathing had gotten shallow. Otherwise, he sat there, traveling beside me in a parallel universe. A slow cosmic music repeating on a loop (that never seemed like mere repetition but more like a staircase headed down, down . . .) guided my journeys with Bob. The shades were drawn. Babaji, Jesus, and other masters stared from posters on the walls. Their gaze was so all-knowing, it little mattered that they were mere paper replicas of faded images. They carried light. This little bit left of them recognized my higher self. Three cats bounded in and out the window and either sat on my legs or brushed by my head (as the process required). They had attended so many rebirthing sessions they knew how and when to help. They were clearly the Egyptian others of California cats.

At the end of the first session, I gave Bob a catalogue of our publishing (North Atlantic Books and Frog, Ltd.). After the third or fourth session he asked me about how we ended up doing Richard Hoagland's *Monuments of Mars* and other books about the Face and City on Mars. As we talked I learned that he was involved in an esoteric tradition that reinterpreted the Face and City in light of ancient and intergalactic events. I thought maybe he could write a book about this version of the Martian mysteries. He came to my house and, after viewing a basketball game together, we sat for hours watching Drunvalo Melchizedek lecture.

I realized that Bob stood at the juncture of two deeply submerged mystery churches—Leonard Orr's rebirthing with its roots in the immortal saints and yogis of India, and Drunvalo Melchizedek's sacred geometry and Pueblo millenary religion with its references to obscure branches of theosophy and uncharted bridges between science-fiction legends and symbolic truths masked in interplanetary tales. Bob stood there, much in the way he practiced rebirthing, as a silent guide, an intentional naif asking to be filled with higher wisdom.

Bob was also in the lineage of Christian Rosenkreutz, who was invited to an alchemical wedding within a ceremony that we ourselves (centuries later) still await the meaning of:

"All on a sudden ariseth so horrible a Tempest, that I imagined no other but that through its mighty force, the Hill whereon my little house was founded, would flye in pieces. . . . Behold it was a fair and glorious Lady, whose garments were all Skye-colour, and curiously (like Heaven) bespangled with golden Stars, in her right Hand she bare a Trumpet of beaten Gold, whereon a Name was ingraven (which I could well read in) but am as yet forbidden to reveal it."

Shaking the hill with her trumpet, she left him a letter. Trembling, he opened it to find words in golden letters on an azure field:

"This day, this day, this, this

The Royal Wedding is."

"As soon as I read this Letter," Christian Rosenkreutz tells us, "I was presently like to have fainted away, all my Hair stood on end, and a cold Sweat trickled down my whole Body."

For this was the famous Wedding, the Wedding of metal and spirit, of time and space, of Alpha and Omega, the ceremony "unless with diligence thou bathe, /The Wedding can't thee harmless save."

Precisely three hundred years later the alchemical wedding is located in the fragile and wounded ecology of the Earth. The tempest is now: pole shift, global warming, ozone-layer deterioration, false prophets. The Lady no longer carries a trumpet or delivers mail but speaks in riddles, through mediums and walk-ins. The Lady serves as proxy for those in the Pleiades, the Thirteenth Dimension, and among colonies we imagine on (or in) our neighboring worlds of Venus and Mars.

Leonard Orr said, "The great immortals are not hiding from us, it is we who are hiding from them."

Leonard Orr also said, "Most people love physical death more than eternal life. . . [But] eternal life is pleasurable. All the immortals I have met are having a great time."

In this book Bob has told his own story in his own words. He makes no attempt to turn millennial mysteries into newspaper facts or self-help instructions. His book is called *Nothing in This Book Is True, But It's Exactly How Things Are* not because the things in it are fictive but because their essential nature is unknowable in simple time and space—not because he is satirizing New Age channeling and apocalypticism (as some reviewers mistakenly thought) but because he understands the irony of our situation trapped between the implacability of doom prophecies and the daily opportunity we have to "rebirth" not only our lives but our past histories, our fates, our loved ones, and in fact the whole cosmos. The story he tells is one of many stories and also one of many versions of one of many stories. It is a story we are living, and it is a story we are changing to such a degree that it is becoming unrecognizable even as it is being absolutely and unambiguously revealed for what it is. That is why nothing is true and yet that same "nothing" is certainly how things are. Anyone who doesn't realize this fundamental paradox of our situation is stuck in one or another prison mythology.

Rebirthing and the Flower of Life are not mythologies; they are acts conferring their own meanings. Accompanied by his feline seers, Bob himself is everyman (everywoman too), a silent spirit-guide to a journey each of us must take ourselves, a voice warning of the terrible things that will happen (but only if we addict ourselves to a vision of the terrible things that will happen).

One day I came in for a session particularly stricken and bereft. I told Bob a sad tale of stuff going wrong in my life. I made it sound totally hopeless. It sounded totally hopeless to me as I was saying it. Then I took a deep breath, stared at Babaji and crew, smiled, and said, "But it's okay."

It wasn't just a line. I felt at that moment utterly pure in my heart and my intention to be.

"It is," Bob said. "It is okay, but only if you say so. Your word must be enough."

So I leave it with readers to decide if this is all true or if none of it is true. Because in the end your own word must be enough. For this lifetime and for all lifetimes and universes to come.

In fact, your word is the only thing that will get you through.

Richard Grossinger is the curator of the Inner Traditions collection Sacred Planet Books and author of *Bottoming Out the Universe* and *Dreamtimes and Thoughtforms.*

Introduction

Has it really been thirty years? In many ways it seems like only yesterday when a guy named Richard first came to me as a breathwork client (I called it rebirthing at the time). Though skeptical at first, he had an unexpected experience in that first session that caused him to sit up and take notice; he then came for twenty-nine additional weekly sessions. During that time a mutual understanding and trust quickly developed between us that led to a most unexpected result.

Yes, I learned that Richard owned a publishing company, and no, I was not auditioning to write a book. In fact, I had internally decided just four years earlier when asked to co-author a book with a breathwork client—and upon learning that she was in fact suggesting that we write a book proposal that we would then use to seek a publisher—that the only way I would ever write a book was if a publisher came to me and asked me to do so. I thought I was safe, but that of course, is exactly what happened.

Prior to this fateful meeting in October 1992, for the previous twenty years I had been an active seeker; I was motivated primarily by an intense desire to heal myself of a crippling low back injury. Upon realizing that everything is a function of consciousness that can only be realized by looking inside, I dedicated myself to exploring the gap between who we think we are, and who we really are. Then I was looking for ways of taking it out of the realm of conceptual under-standing and making it real in my life. Breathwork became my pri-mary tool for doing so; it was a technique that was creating real and lasting healing, and transformation in my life—as well as in the lives of my clients.

Now back to Richard—I quickly learned that his company North Atlantic Books, had published Richard Hoagland's book *The Monuments of Mars*. Since I had recently come across new and essential missing information regarding the unity of life that I knew I needed to be aware of, and since it originally expressed itself in a most unsus-pecting way in the form of other life forms throughout the cosmos, with Mars being one of the centers of attention, I continually pressed Richard for more and more information. And of course, I willingly shared with him all of what I was learning. So one day I received a phone call from Richard asking me if I would consider writing a book on what he termed, "the esoteric meaning of the monuments of Mars." And though my first thought was "No, I'm not an author," I suddenly remembered my decision of four years earlier, and I instantly realized that there were higher forces at work here.

Nothing In This Book Is True, But It's Exactly How Things Are quickly to my complete surprise—and through word of mouth only (not one dime was ever spent on marketing it)—became an overnight sensation. The enormous amount of feedback I've received has been enormously satisfying. As *Nothing* grew in popularity, I was receiving about fifty handwritten letters per week, totaling well into the thou-sands. They were telling me how helpful my book was in the lives of these people. I am grateful to know that I have contributed to so many people in this way.

It has subsequently been published in twenty-five languages; it has sold in excess of five hundred thousand copies, and it continues to be well received to this day. As a result, I have been invited to give my workshops throughout North America, Europe, and Australia.

In this special edition I have kept intact the basic message of the previous edition; I have also expanded upon it by including significant additions and updates, including ten completely new chapters.

I detail from a big-picture perspective, the enormous infusion of higher dimensional energy that is dramatically raising the vibratory rate of the planet and everyone on it. I also give the details of the personal transformation that we must make if we are to survive and thrive, so we can "catch the ride" into higher consciousness in a way that enables Mother Earth to reach critical mass and become "lit from within." This is the story of nothing less than the birth of a new humanity, of the co-creation of Heaven on Earth.

For thirteen thousand years we have been in separation. We see everything as polarized—as good or bad, up or down, hot or cold. We judge everything that happens. The very act of judgment keeps us in this mode of perceiving.

As an example of this separate way of looking at life, if your body gets cold you think of a heater; you think of something outside yourself. This has led to our increasing dependence on technology. And while there have been many amazing advancements in technology, there is also a downside. We have missed the fact that the more technologically advanced we are, the more ignorant we are becoming. We are increasingly separating ourselves from the One Spirit and we are weakening.

We become weaker when we keep giving away our power to technology—that is, to external objects. We then become dependent upon these objects and soon get to the point where we can't do anything for ourselves. We also become habituated to seeking and obeying outside authority.

Oneness does not understand this. It doesn't know what "needing anything" means. Aboriginals and other indigenous people do not know what "need" means; whatever they need just appears because they are in harmony with Nature. The Hopi don't even have words for anything outside themselves. They would speak, for example, of a bird that is perched on a tree branch that is within them.

We have stepped outside this balance, and at this moment, we are learning how to step back in. If we really knew, we could just think the thought "warm" and it is so. We have the capability of changing anything in the reality from within. (I use "the reality" to refer to the outer world, including life situations, circumstances, etc.) Many of us feel that we are just a person with no power and no say regarding how things are in creation. The greater truth is that we can change anything in our environment under certain circumstances—when we are in Unity and not in separation. The Spirit of God can move right through you.

The right brain is our connection to the all that is; it is holistic, intuitive, experiential, and knows only momentary time, *the eternal now*. It is able to intuit the Oneness, like a four-or five-year-old, looking up at the night sky. The left brain sees dots and does not connect them. It is logical, and is locked into linear time, so it is never in the present moment. It needs to see beyond any doubt that separation is an illusion. It must be shown Unity in a series of logical steps as I will do by showing you the holographic universe through the universal language of sacred geometry, and by showing you how the reality that we think we see "out there" is really a holographic projection of our consciousness. When the mind truly sees that there is only One Spirit moving through everything—that there is only one creation pattern, only one law moving through all things—a relaxation occurs; the corpus callosum opens up and communication takes place between both sides of the brain. Things can begin to happen at this point that you yourself would not allow until you were certain that there is only Oneness.

Sixteen thousand years ago we violated galactic law. We were on a very high level of awareness at the time, far beyond where we are now, but by committing a certain illegal act we fell many dimensional levels until we landed in this dense aspect of the reality. We stopped breathing in the manner we formerly utilized, causing the *prana* or life-force energy to bypass the pineal gland, the direct result of which is separation. We now experience ourselves as inside a body looking out at a world that is not us.

There are no accidents, however. In the cosmic scheme of things, this "fall" was necessary in order to allow for a greater possibility. We are now about to leave this place. The whole world is. We have only a short time left before we will no longer be in separation. This is not going to be happening someday, it is happening now.

We live in a time that can be deeply moving, but first, you must be listening with your heart. When you are, you can begin to find a common language that is beyond right and wrong, good and evil.

Your Higher Self is still within you, and it functions well beyond the capabilities of your mind. Reconnecting with this severed aspect of ourselves gives us access to our unlimited potential; where we have reliable access to an infinite supply of inner peace, unconditional love, joy, creative expression, and wisdom. This is the energy of Divine Creator; it is the energy of Source. We must first go down in the process of reconnecting and reclaim our childlike innocence. We must first connect with our Lower Self. The Lower Self is our subconscious mind; it is a child, about two to six years old. Reconnecting and fully reclaiming our childlike innocence is code for the healing work that needs to happen. In the chapters that follow, I will show you exactly how to do that.

Perhaps you're among those who have always felt that there's much more to our existence than we have been led to believe; and you've always felt that we're headed somewhere. But you may not know that the speed at which we're now evolving is unheard of in the universe, thanks to a wildly successful interplanetary experiment conducted in 1972. These adventures in other-dimensional reality are coming to fruition on Earth right now!

You will be guided through the ancient Mystery Schools—the Naacal Mystery School in Lemuria and Atlantis, and then onto the two twelve-year Egyptian Schools, culminating with the final imitation in the Great Pyramid and the ascent into Christ-consciousness.

Around the circles and spirals of sacred geometry, in and out of the magic of Circular Breathing and the activation of the light-body, even surviving the potentially cataclysmic pole shifts intact, we will eventually assemble internally our own Merkaba, the universal pattern of creation, and join the ascended masters.

We will meet the forces working to advance our evolution of consciousness, and those seeking to block it. When we reach critical mass, Mother Earth will shift into a higher dimensional level. Many galactic presences have already gathered around our planet, watching and waiting to observe this unprecedented event.

This is a time of great celebration as we move out of the darkness and into the light. It means that the veils will be lifted, we will remember and live our intimate connection to all life, we will be allowed to reunite with our cosmic brothers, and to move about the universe. We will completely redefine what it means to be human!

1
First Contact

Everyone is on a journey of some sort. Everyone has questions about the meaning of life, and we listen whenever a small part of this quest is undertaken and new information is revealed.

It all began for me back in 1972, when I discovered at the suggestion of a friend, a classic piece of literature called *The Book: On the Taboo against Knowing Who You* Are by Alan Watts. Since my overt aim was to become a better bowler—I was searching for a way to calm and center myself so I could enhance my performance in pressure situations—I'm not sure why my friend recommended this to me, but it hardly matters. What this book did was rekindle a spark within me, a deep inner resonance. It reminded me of the sense of awe and wonder I had as a small child when life—for its own sake—was a truly exciting experience!

As a result of my growing realization that we're much more than just our limited identity, I began soaking up as much information as I could get. I read a number of books by Watts and other authors, I learned Transcendental Meditation (my bowling game improved dramatically), I took the most popular seminar of the day, the two weekend *est* training, and then I immersed myself in a conscious breathing technique known as rebirthing. With all this input, and with the aid of numerous teachers, I experienced a quantum leap in personal growth; I was awakening from a deep slumber. And because I was motivated by a deep desire to learn, I felt quite confident that whatever I needed to know would appear whenever I needed to know it. And that is exactly what has happened.

This then, is the story of my continuing personal journey of discovery, of how I reached a turning point after twelve years of emotional body training when I began to realize there was indeed missing information that I needed to be aware of. It made no difference that I had no idea what the particulars could be, let alone when and how they would begin to reveal themselves. I knew I already had the necessary ingredients: one hundred percent intention, along with a willingness to follow the Spirit within without hesitation.

As a result, my trip has been magical in every way. I have been "guided" to be in the right places at the right times to meet the perfect people who have given me exactly what I needed to stay on the path. This has proven to be a never-ending journey.

It began innocently enough in January 1991, when I was leading a week-long training at Campbell Hot Springs (at that time a rebirthing center of Leonard Orr's) near Sierraville, California. Leonard Orr is the founder of a breathwork process that he termed rebirthing. I spent most of 1980 training with him, and after moving through seven years of "speed bumps", have been a professional teacher of conscious breathing ever since.

Though my initial motivation was out of necessity—I was looking for a way to heal a low back injury—breathwork quickly developed a life of its own for me. For more than forty years now, I have refined and reinvented it, and in its' current incarnation, it is a technique that I call Breath Alchemy.

In simplest terms, Breath Alchemy is a tool for helping you in discovering and mastering your innate ability to transform your fears and limiting beliefs into powerful, purposeful life-enhancing guidance systems, so you can be truly present, joyful, and peaceful in all areas of your life. More about that later.

I always enjoyed my time at Campbell Hot Springs. It covers six hundred and eighty acres of pure Sierra wilderness located right in the middle of nowhere. Just my kind of place.

In addition, there was always something special, something magical about the trainings, and this week was no exception. So, if that was all that happened, I would have been content with my stay there and departed looking forward to my next visit.

Then I met Doug!

Just prior to the conclusion of the week's events on Friday evening, I glanced at a book he left in the lunchroom called *Space-Gate: The Veil Removed* by Gyeorgos Ceres Hatonn. It was full of fascinating, frightening, unbelievable information about UFOs, aliens, cover-ups, and conspiracies. I had only a short while to glance through it; then it was time for the seminar.

As soon as the class concluded, there was Doug standing in the hallway, almost as if he was waiting for me. I told him what I had seen and asked for an explanation. We talked for hours in a hallway that unlike the seminar room was not heated. And yes, it does get cold at night at 6000 feet in the Sierra in January! We'll come back to our conversation in a moment, but first let me fill you in with a bit of background information.

The beginnings of the modern UFO era are often traced to an Idaho pilot and businessman named Kenneth Arnold who witnessed nine UFOs in June 1947, while flying over the state of Washington. He compared the flight of these unknown objects to a saucer skipping across water, which resulted in the creation of the term "flying saucers." Arnold's experience was one of almost nine hundred sightings across the U.S. that summer; this kicked off a nationwide UFO frenzy.

What neither Arnold or the rest of the world knew, was that the U.S. military was already busy with secret full-scale investigations of the phenomena, quite likely the direct result of the many UFO sightings by military personnel toward the end of World War II. "Foo-Fighters" they were called. We thought they were secret weapons of the Germans; the Germans thought they were ours.

As early as 1948, before the cover-up was fully implemented, top-secret studies by the military and the FBI, had concluded that UFOs were real, that they were metallic disc-shaped craft, capable of great speeds and evasive maneuvers, and of unknown origin. Subsequent studies concluded that the craft were, most likely extraterrestrial. Of course, the public was not let in on these conclusions; in fact, the existence of these studies was kept secret until 1976.

Now that we have a bit of a background, let's get back to Doug. Our exchange was truly an amazing experience; he answered all of my questions with a great deal of patience and understanding. And as improbable as it seemed, I now knew that the next phase of my journey had begun. What it would eventually lead to, far surpassed even my wildest and most fanciful dreams! Then as we finally parted for the evening, he gave me the book and a manuscript written by Bill Cooper.

Bill Cooper is now honorably discharged from the navy. He states that in 1972, he saw two reports relating to government involvement with alien creatures while working as a quartermaster with an intelligence briefing team for U.S. Admiral Bernard A. Clarey, then Commander in Chief of the Pacific Fleet. Bill served in that capacity from 1970 to 1973.

He states that the two reports he saw were: 1) Project GRUDGE/ BLUE BOOK Report No. 13, and 2) MAJORITY Briefing.

The GRUDGE report he saw contained about twenty-five black and white photographs of "alien life forms" and information about them. The MAJORITY report had no photographs, Bill said, but contained information about the government's growing concern with the alien interference on this planet.

Doug also said that he had a number of videotapes and would be willing to show them to me. I stayed up most of that night reading and then spent much of the next day watching these videos.

To summarize what I learned: there was a great deal of UFO activity in the late 1940s, most of it in the New Mexico area. This included numerous incidents of downed or crashed alien craft, the most famous being the July 1947 recovery of a crashed UFO on a ranch near Roswell, New Mexico, along with three or four alien bodies. At the time, Roswell was the home of the Roswell Army Air Field of the Eighth Air Force, which housed the 509th Bomb Group, the only squadron in the world entrusted with atomic weapons.

Roswell residents on the fateful evening of July 2 saw a bright object streak across the sky. It exploded about seventy miles outside of town, scattering debris over a large area. Rancher William "Mac" Brazel found the strange metallic fragments, and days later, he reported the matter to authorities. Roswell intelligence officer Major Jesse Marcel was dispatched to the scene to collect the wreckage. Marcel was certain that it was not a weather balloon, nor an aircraft.

He said the foil, no thicker than that in a pack of cigarettes, was virtually indestructible. It was, in his words, "not of this Earth."[1]

Base commander Col. William H. Blanchard called in press liaison Lt. Walter Haut, who issued the following press release:

> The many rumors regarding the flying disc became a reality yesterday when the intelligence office of the 509th Bomb Group of the Eighth Air Force, Roswell Army Air Field, was fortunate enough to gain possession of a disc through the cooperation of one of the local ranchers and the sheriff's office of Chaves County. The flying object landed on a ranch near Roswell sometime last week. Not having phone facilities, the rancher stored the disc until such time as he was able to contact the sheriff's office, who in turn notified Maj. Jesse A. Marcel of the 509th Bomb Group Intelligence Office. Action was immediately taken and the disc was picked up at the rancher's home. It was inspected at the Roswell Army Air Field and subsequently loaned by Major Marcel to higher headquarters.[2]

When the local newspaper tried to put this out, the wire service was interrupted. A message came across to the local news service stating that they were to cease transmission immediately, that what they were transmitting concerned the national security. The message was from the FBI.

Major Marcel was told to load the wreckage onto a B29 and fly it to Eighth Air Force headquarters in Fort Worth, Texas, where General Roger M. Ramey took control of the debris. He ordered Marcel and others to keep quiet and issued another press release saying that the incident was nothing more than a crashed weather balloon. As he and Marcel posed for reporters, the real wreckage was flown to Ohio under armed guard.

Wright Patterson Air Force Base in Ohio was the original center for all material of this type to be brought together. It was headquarters for the foreign technologies division of the U.S. Air Force. It later became headquarters for the alien technology division of the USAF.

Then two craft were found in February and March of 1948 near Aztec, New Mexico. Seventeen alien bodies were recovered, along with a large number of human body parts. This immediately lifted the security lid to Above Top Secret.

The next significant occurrence was in 1949, and again it was near Roswell. Two discs crashed and five alien bodies were recovered. In combing the area to make sure that all the wreckage was picked up, a live alien was found wandering in the desert. He was captured and named EBE (extraterrestrial biological entity). EBE was taken to Las Alamos where he was placed in a Faraday shielded environment. Why? Because he had the ability to read the thoughts and transmit thoughts to people on the outside. He also had the ability through a device that he wore around his waist to literally walk through walls.

EBE became ill and there was an unsuccessful attempt to try and save him. In the attempt to save him and also in an attempt to gain favor with this technologically superior race, they began broadcasting into space for help. They hoped to make friendly contact so the aliens would see us as a benevolent people trying to help one of theirs. We realized that in any military confrontation, we would have had no chance whatsoever.

EBE died in June 1952, and as of yet, there was no answer from our attempts to communicate. The movie *E.T. The Extraterrestrial* was the thinly veiled story of EBE. In real life, it was an air force officer, not a little boy, who developed a close relationship with EBE. When EBE died the officer was deeply hurt; he had lost a friend. In the movie, E.T. becomes very ill and the boy plays the role of the air force officer, but in the film, he lived. E.T. call home was *Project Sigma*, the attempt to establish contact with the aliens.

By the way, in a special White House screening of the movie, President Reagan whispered in Steven Spielberg's ear, "You know, there are fewer than six people in this room who know the real truth of the matter."

In 1953, Dwight D. Eisenhower's first year in office, at least ten more crashed discs were recovered, along with twenty-six dead and four live aliens. In early 1953, the new president turned to friend and fellow Council on Foreign Relations (CFR) member Nelson Rockefeller for help. They began planning the secret structure of alien-task supervision, which was to become reality within one year. The idea for MJ-12 was born.

Majesty Twelve (MJ-12), also known as Majority Twelve, was created early in 1954 to oversee and conduct all covert activities concerning the alien question. It consisted, according to Bill Cooper, of Rockefeller, Director of Central Intelligence Allen Welsh Dulles, Secretary of State John Foster Dulles, Secretary of Defense Charles E. Wilson, Chairman of the Joint Chiefs of Staff Admiral Arthur W. Radford, Director of the Federal Bureau of Investigation J. Edgar Hoover, six men from the executive committee of the Council on Foreign Relations known as the "Wise Men," six men from the executive committee of the JASON

Group (a secret group formed during the Manhattan Project), and Dr. Edward Teller.[3]

No order could be given from the nineteen-member group, and no action could be taken without a majority vote of twelve in favor, thus "Majority Twelve."

It is significant that Eisenhower, as well as the first six MJ-12 members from the government, were also CFR members. This gave control of the most secret and powerful group in government to a special-interest club that was itself controlled by the Illuminati.[4]

A contingency plan was formulated by MJ-12 to deceive anyone who came close to the truth. It was known as Majestic Twelve (MJ-12). It is a fraud, and it has successfully thrown most researchers off the trail.

The next alleged major event occurred in 1954, when our government made contact with a race of aliens that has since become known as the "Greys." Evidently they landed at Edwards Air Force Base, met with President Eisenhower, and signed a formal treaty.

According to Bill Cooper:

> The treaty stated: The aliens would not interfere in our affairs and we would not interfere in theirs. We would keep their presence on Earth a secret. They would furnish us with advanced technology and would help us in our technological development. They would not make any treaty with any other Earth nation. They could abduct humans on a limited and periodic basis for the purpose of medical examination and monitoring of our development, with the stipulation that the humans would not be harmed, would be returned to their point of abduction, that the humans would have no memory of the event, and that the Alien nation would furnish MJ-12 with a list of all human contacts and abductees on a regularly scheduled basis. . . .
>
> It was also agreed that bases would be constructed underground for the use of the Alien nation, and that two bases would be

constructed for the joint use of the Alien nation and the United States Government. Exchange of technology would take place in the jointly occupied bases.[5]

Shortly thereafter, two outcomes became clear: 1) The Greys had ignored the terms of the treaty by abducting far more humans than they said they would. They also carried out mutilations, both human and animal. The Greys said that this was necessary for their survival, that they were a dying race and that their genetic material had deteriorated to the point where they were no longer able to reproduce. They said they needed our genetic material or they would be history. 2) Our weapons were no match for theirs. As a result, it was necessary to remain on a friendly basis with the aliens, at least until we could develop weapons systems that might combat them. Of course, an "above top-secret" security lid was put on this, with government agencies formed to investigate.

The above barely scratches the surface. There was much more information, all equally mind-blowing, as if straight out of a science-fiction book. With all the Ace paperback qualities to the story, it would have been all too easy to laugh it off as the product of someone's imagination, or the hoax of the century—and that certainly was the prevailing attitude of the day.

But for me, at the risk of being labeled a "kook" or even worse, one of those "conspiracy theory nuts," it fit, like the missing piece of a puzzle. It answered a lot of questions; it felt right intuitively. One of the many results of my many years of breathwork was a highly developed intuition. My intuition said "yes" to this and I trusted it.

Reality Check

I went back home bursting with this new information. I couldn't wait to tell anyone and everyone who would listen. I could tell you the boring details, but I won't. I could tell you how it cost me a client or two and probably a friendship or two. I could tell you how, much to my surprise, almost no one wanted to hear it. Almost—I did find a few exceptions.

I learn quickly. I soon became very careful about volunteering this information, and I even began to question it myself.

I was no stranger to the possibility that what passes for truth in the mainstream media is in fact a thinly disguised veil for the real working of governments. But could it really have gone this far? Were there really little grey aliens behind the scenes giving us *Star Trek*-type technology in exchange for abducting some of us and genetically experimenting on those unfortunates? Could I really trust my intuition this far, or had I gone off the deep end?

I went to the Whole Life Expo in San Francisco in April and heard a presentation by two of the more moderate researchers, Stanton Friedman and William Moore. They both agreed about the Roswell incident; in fact, both had researched it *ad nauseam*. However, Moore in particular went out of his way to paint Bill Cooper as a kook who could not be trusted. Was I convinced? No. Did it open up some new doubts? Yes.

I also picked up a video entitled *Hoagland's Mars: The NASA-Cydonia Briefings* and was introduced to this most interesting study regarding alleged "monuments" on Mars. Why hadn't the media reported this?

I pretty much closed up shop on all of this after the Expo. There was nowhere to go with it. Besides, I had a life to live and this was getting me nowhere fast! I did, however, connect with a group of like-minded people who met on a monthly basis. I attended a few of their meetings and received their regular mailings. In June of 1992, I received a fateful letter in the mail (fig. 1.1).

As I gazed at this letter, I was instantly engaged. It spoke of ascension and breathing as being the keys to higher consciousness. I was especially intrigued by the mention of the ascended master named Thoth,

DYNAMICS OF HUMAN BEHAVIOR

MADELYN BURLEY-ALLEN
FOUNDER

June 1, 1992

Greetings!

On Sunday, June 7th, I will be showing the first of five videos of Drunvalo Melchizedak's Dallas, Texas workshop. Each video is approximately 5 hours.

TIME: Sunday, June 7 from 1:00 - 6:00 p.m.
LOCATION: 1710 South Amphlett Blvd.
            Conference Room 126 on the first floor
            This is an office complex that is North
            of the Dunfey Hotel (Map enclosed.)

Please RSVP by Friday, June 5 as it is important for me to know who is coming and their phone number. If I have not heard from you by the 5th, I will assume you are not coming.

The following describes the focus of these videos:

"Resurrection, moving consciously into the next vibratory dimension can and must be experienced and lived if we are to survive into the 21st century. Our planet, whether you are aware of it or not, is already deep into the transformation. The ascension process begins when a human remembers his or her crystal energy field. There is a field of energy that is fifty-five feet around the body that is geometrical and crystalline in nature. The remembering of this field is triggered by a series of metaphysical drawings that are light replicas of the 44 + 2 chromosomes in every cell of your body and specifically in your pineal gland. This will activate a higher purpose of the pineal gland which is to allow a forgotten ancient way of breathing to return.
        This breathing is a key to higher consciousness and dimensional translation. By simply breathing in a different way than we do now, and by directing the pranic flow through the human crystal field, a new world will literally open unto you. This breathing will allow you to make direct contact with your higher Self so that trusted and clear guidance can come from within. It will give you unparalleled protection while you are in the ascension process. It will give you a means to heal yourself and later others. This breathing will allow you to remember who you are and your intimate connection with God.
        This video workshop is presented by Drunvalo Melchizedek. The teaching itself comes from Alpha & Omega, order of Melchizedek and from Thoth, the Egyptian (Atlantean), also known as Hermes of Greece, who resurrected long ago and was an immortal that was physically living on earth until a few months ago."

I look forward to your participation. This information is extremely significant for your ascension process.

Love and light,

Madelyn Burley Allen
Madelyn Burley-Allen

MBA:jb
Enclosure

Figure 1.1

whom I would soon discover has been alive and in the same body for fifty-two thousand years. It looked too good—this was exactly what I was learning in breathwork!

Even though it seemed like an enormous commitment to sit and watch twenty-five hours of video (it actually turned out to be thirty-two hours), I responded with an immediate yes! I had no idea what to expect; I just knew that I wanted to go. The videos were part of a four-day workshop entitled *The Flower of Life*, led by Drunvalo Melchizedek. Was this a real person? What kind of a name was this?

There was no formal workshop—it was just a group, about twenty of us, who gathered to watch the videos. I recall sitting in the back of a fairly long, rectangular conference room. I was too far away to see much on the TV, but I could hear, and what I was hearing was rather amazing. I walked up the side aisle to see what this guy looked like. Drunvalo was obviously the possessor of great knowledge, and he was presenting it in such a sincere, innocent, and humble way. He didn't have anything extra going on with it. He definitely wasn't trying to prove anything or convince anyone of anything. In addition, his material was all-inclusive—he was talking about Earth changes, the environmental problems, the Greys, the secret government—all in the context of the bigger picture. He was in full possession of an awareness that something bigger than all of this was going on, and over the course of thirty-two hours, he slowly let it unfold. It was this broad approach that impressed me the most.

I was so impressed that I purchased the videos, and they were my preferred source of entertainment for many weeks to come. In fact, they were my *only* source of entertainment; I devoured them as though I was preparing for a series of final exams, and the beauty of it was that my motivation was pure; I felt like a kid in a toy store.

I began to realize that this indeed, was the "missing information" that I was looking for. Yes, the input I received from Doug was very important for my awakening, but it was now becoming clear to me, that it was only in preparation for this, the main course. It fostered a much deeper inner awakening that quickly became a wildfire, fueled by the fire of purpose that continues to burn brightly to this day.

I learned that Drunvalo was available for telephone conversations on Saturdays, so I began to call him on a regular basis. I found these weekly talks to be of enormous value; they served to clear up any confusion or misunderstandings with regards to the material. They propelled me into "warp speed."

He also told me that he was phasing out of giving this workshop, and that he was looking for people to take it over for him. I'm sure he noticed my deep level of commitment, that in fact, I was determined to master this material to match his level of understanding. So, he asked me if I was interested, and indeed I was!

I then spent nine days in Austin, Texas, and was trained by him to give this workshop. I took my training and ran with it. I proceeded to give the *Flower of Life* course for the next twenty years to thousands of eager students all over North America, Australia, and Europe.

So, what did Drunvalo have to say that was so astonishing? Well, for one, the letter's mention of breathing and ascension is what got me there, and I wasn't disappointed.

But how did that relate to UFOs and little grey aliens?

Very good questions, so stay tuned. But first we need to look at a pair of additional riddles.

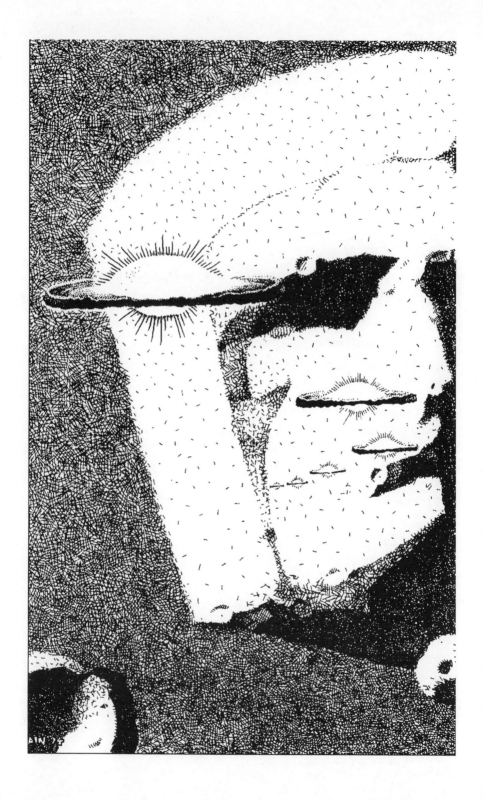

2
What's Going On?

Doug also gave me copies of the many videos we had watched. Most of them were multigenerational copies, some of which took a great deal of desire and concentration to watch. One of them was a series of reports produced by Las Vegas TV station KLAS, in which one Bob Lazar claimed that Americans were working on alien craft at a Nevada test site called Area 51.

Lazar claims that he was hired by the navy as an engineer on a top-secret project. He was under the impression that he was going to be working on an advanced propulsion system.

He says he was flown from Las Vegas to Groom Lake, put on a windowless bus, and driven about ten miles south to Area S-4 at the Papoose Lake dry lake bed. This is reportedly the site of nine heavily camouflaged hangars alleged to contain an equal number of unconventional aircraft. He said the hangar door was open as they drove up to one of them, and inside was a flying saucer.

He didn't fully realize that it was of alien origin until sometime later when he was taken inside the craft. Upon seeing the diminutive size of everything, Lazar was quite convinced that it was not made for humans.

The true motivation for the government to have hired him then emerged: this was a reverse engineering project—the process by which an advanced system may be used as a learning model for creating another.

Lazar said the disc was constructed with three levels: the upper level, which he never saw; the middle level, where you enter; and the lower one, where the gravity amplifiers hang down and you can access the bottom part of the craft.

He said there were no obvious control panels; there were only three chairs, the reactor itself, and the gravity amplifiers. Underneath them were the other components of the propulsion system.

Lazar claims that his first few days were spent under guard, reading briefing documents containing detailed knowledge of the aliens. These beings were the Greys, said to originate from Reticulum 4, a constellation in a star system in Zeta Reticuli. "Reticulum 4" referred to the fourth planet out from that sun.

Lazar's job, in a team of two, involved working with the power and propulsion system of the craft. The power source, according to Lazar, was an antimatter reactor, a total annihilation system, reacting matter and antimatter, which is one hundred percent efficient in its conversion of matter to energy. It was small enough to hold in your hand, yet it put out more power than our full-sized nuclear power plants.

This allowed the ships to produce their own gravitational fields. According to Lazar:

Gravity distorts time and space, just like if you had a water bed and put a bowling ball in the middle. It warps it down—that's exactly what happens to space. Imagine you were in a spacecraft that could exert a tremendous gravitational field by itself. You could sit on any particular place, turn on the gravity generator, and actually warp space and time and fold it. By shutting it off you would be a tremendous distance from where you were, but time would not have even moved . . . when you harness gravity you harness everything. It's the missing piece in physics.[1]

Lazar says the technology to harness gravity not only exists but is being tested at S-4.

Element 115 is the fuel for the reactors. By bombarding 115, antimatter is produced. Not found on the periodic chart, element 115 can be stored in lead casings. Lazar says the government has five hundred pounds of it, and it cannot be made here. "It has to come from a place where super heavy elements could have been produced naturally, perhaps next to a much larger sun or a binary star system where there would be greater mass."[2]

Lazar says he was fired after making a blatant security breach. He brought five friends to the outer perimeter of the base to witness the test flights. As time went on, they became lax and eventually got caught.

He was debriefed at an air force base north of Las Vegas. Weapons were pointed at him, his life was threatened, and personal records were erased, making him a virtual non-person. At that point Lazar decided to go public as an insurance policy on his life. He went to station KLAS in Las Vegas and began a series of interviews with investigative reporter George Knapp.

Lazar is not the only one to claim inside knowledge of the flying discs at the test site; he is, however, the only person to say so publicly.

Knapp has communicated with several people who say they know of the saucer program:

A technician in a highly sensitive position told him, "It is common knowledge among those with high security clearances that recovered alien discs are stored at the Nevada test site."

A Las Vegas professional who while in the military was stationed at the test site says that he saw a flying disc land near Area 51, that it was quickly surrounded by security personnel, and that he was taken away and debriefed for several hours.

An airman who worked at a radar installation says that he and his fellow servicemen watched UFOs flying over the Groom Mountains over a period of five nights. Radar indicated the UFOs were traveling at speeds of up to 7,000 miles per hour and then would stop on a dime. He was finally ordered to turn off his sensors for that area and to keep quiet because what he saw "did not happen."

By the way, it is alleged that a program called Project Aquarius, established in 1972 to fly recovered alien spacecraft, still exists to this day.[3]

In 1997, Col. Phillip J. Corso (Ret.) came forth to reveal that reverse engineering was indeed a common occurrence with regard to downed alien craft. Corso, in his book *The Day after Roswell,* said he led a double life in the military: On the outside he researched and evaluated weapons systems for the Army, while deep inside the Pentagon he was responsible for the Roswell file, the army's darkest secret.

He said the military seeded the industrial complex through reverse engineering, with components discovered in the wreckage at Roswell that led to today's integrated circuit chips, fiber optics, lasers, and stealth technology.

Two of the videos from Doug were entitled *The Pleiadian Connection* and *Contact.* They told the story of the contacts of Edward Albert "Billy" Meier, with a group of beings from the Pleiades. Included in this presentation were a number of photos of Pleiadian "Beamships" taken by Meier.

Meier, a one-armed Swiss farmer with a sixth-grade education, says that his contacts began telepathically at the age of five, and after eleven years of these communications, he had more than a hundred and thirty visitations from these beings between January 1975 and 1986.

Meier isn't the only person to claim extraterrestrial visitations, but he is the first to document his contacts in such stunning detail, with more than one thousand photographs. These are clear daylight pictures of multiple flying discs (two or more ships in a photo is almost unheard of in UFO circles), with identifiable reference points in both the foreground and background. They were all taken by an Olympus 35mm camera with a little wheel on the back to advance the film.

Meier also had 8mm film footage, sound recordings, about seventy eyewitnesses, metal samples that were representative of the type the ships were made of, and reputed landing sites where the grass was pressed and swirled in a counterclockwise direction that, one year later, still had not sprung back.

Now in case you're wondering how it's done, just how do these guys successfully traverse a distance of five hundred light years? Good question, is it even possible? Well, here is an interesting explanation from *The Pleiadian Connection:*

The ship speeds up to approximately the speed of light. At that point the mass speed correlation (the energy of the universe pressing in on the ship) is tremendous. The ships are protected by an energy screen that holds that off. Just at the right moment they lower those screens and all that energy is forced in against the ship. They use that energy in kind of a compression fashion to do something really unique—they convert the ship and all matter involved with it into thought. They call it spiritual energy; we would think of it as thought. It no longer exists in the material frame. That allows them to step outside of time into "null time" and then move that thought at a much higher speed to its destination. Then the ship rematerializes, allowing them to traverse vast distances literally at the speed of thought. It took three and a half hours to get up enough speed to make this conversion, then the jump took literally no time. Then it took three and a half hours to slow down to enter our solar system. Why three and a half hours? Because they need to be one hundred and fifty-three million kilometers away from the nearest orbiting body before they convert from null time to time. The conversion causes a rip in time. And any planet that may be too close could get pulled out of its orbit. So the "jump" must be made outside of solar systems. So, they were one hundred and fifty-three million kilometers away, and it took three and a half hours to cruise through our solar system.[4]

A five-year investigation into Meier's allegations began in the spring of 1978.[5] It was headed by Wendelle Stevens, a retired air force officer who for twenty years had been one of the world's leading UFO researchers. This was a private investigation, since governments do not investigate what, according to them, does not exist.

The team visited the sites where the pictures were taken, and the photo scene was reconstructed for computer comparison. The investigators were impressed by the vastness of the valley, the steep drops, and the obvious difficulty of hanging models from anywhere to fake the pictures.[6] Computer analysis of the photos by NASA's Jet Propulsion Laboratory and two universities found no evidence of a hoax.

An IBM chemist concluded that even though most of the ingredients of the strange metal could be identified, the manufacturing process could not. It was beyond our technology and clearly not of this Earth. He claimed that the metal had to have been made in some sort of vacuum process.

The team showed the Meier material to a variety of experts in an attempt to learn if it could be duplicated. It was estimated that it would require expertise in at least a dozen different fields, and that it would take at least ten thousand man-hours and more than one million dollars to duplicate. Even then, computers could still spot any fabrication.

The bottom line: outside experts concluded that Meier's story is legitimate. Yet both Friedman and Moore jumped all over the case in their talks at the Whole Life Expo, calling Meier the hoaxer of the century.

So, was Billy Meier really being visited by the Pleiadians, or was this just his idea of a joke? And what about Bob Lazar? Is he telling the truth? If not, he certainly has an imagination. Or maybe he is one of a number of government agents purposely putting out disinformation—part of which may be true, and part of which is obviously false, designed specifically to lead you in the wrong direction. And if it is disinformation, what are they really hiding?

Those videos, coupled with the Cooper and *Space-Gate* material, led me to the inevitable question: What's going on here?

And speaking of what's going on, what about one of the more interesting controversies of our time? Yup you guessed it—did we really land on the moon?

Did you see the 1978 thriller *Capricorn One*, the story of our first manned mission to Mars? Sounds exciting right? Well wait a minute, just mere seconds prior to launch, the project director had a rather unexpected surprise for the would-be Martian visitors. It turns out that the life-support system was faulty and that NASA couldn't afford the publicity of a scratched mission. The plan then was to fake the Mars landing in a movie studio at a remote base and keep the astronauts there until the mission was over. So, the three astronauts, one of whom was played by O.J. Simpson, were pulled from the capsule and the rocket left Earth unmanned. And it almost worked.

Let's take a look at the Apollo Missions; why would we fake them in the first place? In recalling the history of the time, at the conclusion of World War II, the United States acquired the majority of the German rocket scientists, along with most of Germany's related information and hardware. The confiscated information included plans for Earth satellites as well as multistage rockets that would eventually be fired at North America.

So, if this was in the German plans, and since our newest enemy the U.S.S.R., also obtained German scientists, and pursued the development of rockets with a great deal of enthusiasm, it was concluded that this weaponry might indeed be used against us someday. But why? And why would the Soviets fear us at all; after all we were allies in the recently concluded global conflict. Well remember not only did we have the bomb well before them, we also used it! Could it be that the two horrific explosions over Hiroshima and Nagasaki in August 1945 were intended more for the Russians—the first shots fired in the Cold War, rather than for an already thoroughly defeated Japan?

Add to this the fact that in 1953, it was discovered that the Soviets were well along in this endeavor and had already designed a rocket capable of carrying their atomic bombs. In the era following World War II and extending for decades beyond, we were in the height of the Cold War; we were locked up in a nuclear arms race with no end in sight, and the space race was merely an extension of that.

The technology necessary to launch the massive Saturn 5 rocket and an intercontinental ballistic missile was virtually the same. When the Soviet Union launched mankind's first satellite in 1957 Sputnik 1, there was grave concern that they had mastered space ahead of the United States and might use this advantage to launch a first nuclear strike from orbit.

When they also put the first animal in space, then the first man in space, then achieved the first spacewalk, then the first crew of three, and then the first ever of two simultaneously orbiting spacecraft, concern turned to fear and then horror as America watched their communist enemy achieve all these firsts with no hope in sight of ever catching up. We were on red alert!

Then President Kennedy issued an irreversible ultimatum of putting a man on the moon and returning him safely before the end of the decade. This was in spite of the fact that one miserable failure followed another; and our only manned success to date was a sixteen-minute flight that didn't even achieve orbit by astronaut Allan Sheppard. We had some catching up to do indeed.

So did we really land on the moon on July 20, 1969? Well consider the following:

According to William Kaysing, a six-year NASA contractor for the

Apollo missions, a classified interdepartmental memo rated the odds of a survivable manned lunar mission on its first attempt with 1960's technology, at only one in ten thousand.[7]

Not very good odds, right? Well consider the fact that beginning one thousand miles above the Earth and extending upwards for twenty-five thousand miles, lies a lethal band of radiation known as the Van Allen belt. In order to survive the ninety-minute journey through this radiation field, necessary to reach the moon and return, six feet of solid lead shielding between the astronauts and the exposure outside would be required. The Saturn rocket, used by Apollo, was three hundred sixty-three feet tall, and weighed 6,400,000 pounds. To add additional tonnage in the form of a lead barrier completely surrounding the crew members, would have made it impossible for the vehicle to get off the ground.

Is it any wonder then that every space flight has maintained an altitude well below this one-thousand-mile lethal barrier? All that is, except the Apollo missions. Maybe that's why the Soviets, even though having spent many more manned hours in space than the Americans, only sent an unmanned probe to the moon. The Apollo spacecraft's narrowest shielding was less than 1/8 inch of light aluminum; and the question must be asked, "Was this sufficient shielding to protect the crew from this extremely dangerous twenty-five-thousand-mile-thick band of radiation?"

In 1998, the space shuttle flew to its highest altitude ever of four hundred miles; yet it was still hundreds of miles below the Van Allen belt. And make no mistake, the radiation was so severe that the astronauts inside of the shielded spacecraft suffered from radiation poisoning according to an official NASA report. In fact they were considerably ill, enough to be quarantined for some time.

So how did Apollo make it through the belt when the space shuttle couldn't get to within six hundred miles of it without encountering severe problems?

As a result of this, CNN issued the following report:

The radiation belt surrounding Earth may be more dangerous for space-walking astronauts than previously believed. Scientists say the phenomena known as the Van Allen belt can form killer electrons when the Earth's magnetic field changes. These electrons that are being studied could have an important effect, not only on satellites which has happened in the past, but could also affect the astronauts by creating large doses of radiation that could influence their health. The electrons can penetrate through various materials including space suits, and can pass through in fact, the walls of the space station, and can create high charges deep inside of these objects.[8]

Astronaut Alan Bean during the Apollo 12 Mission, allegedly walked on the lunar surface; but he didn't know about the whereabouts of the belt. When asked if he experienced any ill effects from the belt, he replied "No, I'm not sure if we went far enough out to encounter the radiation belt. Maybe we did, I don't know the distance to the Van Allen radiation, and if we did, it wasn't a problem."[9]

Then there's the strange case of no burn crater underneath the lunar modules ten-thousand-pound engine despite the fact that during ground tests there was a real concern for the vehicle falling into the hole the engine created as it descended. As a result, all subsequent flights had to have the same discrepancy, which was explained away by the effects of no atmosphere. And there isn't any dust or dirt on the landing pod feet, as if it was just set down gently onto a stage.

There is also the question of different shadow angles. When all the object are lit solely by the sun as all the scenes on the moon were said to be, then all shadows will run parallel with one another. Yet there are dozens of photos showing shadows of astronauts, flags, rocks, and other objects falling in different directions up to ninety degrees apart; this is impossible without secondary lighting.

There are many pictures which show moon rovers with no wheel tracks in front or behind them (as though they were set down into place) even though there are many footprints all around. There are pictures of astronauts shown with footprints all around them, but no prints leading to or from where they are.

Around 1990 and years behind schedule, the same space program that successfully sent astronauts to the moon and back, couldn't put into Earth orbit a telescope with a lens that focused. And yet two decades earlier, a mission one hundred times more complicated, worked on its first and on every subsequent occasion.

There's more, but this is probably enough to open up the question of whether we landed on the moon or not. Even though the answer is yes, it almost certainly was not accomplished through the Apollo program. Maybe this is why the returning crew looked dejected, rather than triumphant, at their press conference. Could it be that they were forced into lying about the alleged greatest ever accomplishment of mankind?

And oh yes, in spite of a public display of antagonism between the U.S.S.R. and the U.S.A., there was something else going on beneath the surface, something else indeed! I'll have more on all of this later, so stay tuned.

I could cite additional riddles. What about the crop circles? The media back in 1991, trotted out a couple of jokers named Doug and Dave, who claimed full responsibility for the whole thing. They even showed us how they did it; well that ought to solve the mystery once and forever! Funny thing though, at the very minute Doug and Dave were "confessing," crop circles were being formed in Canada. It must have been the wind (fig. 2.1).

According to Colin Andrews, a well-known "crop circle" researcher, more money was spent on a TV documentary "proving" that Doug and Dave did it than legitimate researchers had to spend in their many years of study. Why?

Figure 2.1.
As we shall soon see, a most significant image is represented
in this crop circle. Photo by George Wingfield.

It almost doesn't matter if any of this is true or not. Just the fact that all this information is falling around us, for whatever reason, is a clear indication that we have passed into a strange new epoch.

Obviously, something is going on here on Planet Earth that is not normal, whether it is the thing being discussed or the thing it masks. As a civilization we gathered a certain amount of information since the days of Sumeria six thousand years ago until 1900. Between 1900 and around 1950 we doubled that amount of information. Then from approximately 1950 to 1970 we doubled it again, and from 1970 to 1980 again.

Information has been doubling so rapidly that as of 1992, NASA was about eight years behind in getting all of this data into their computers so it could even be used. We can only imagine how quickly all this information is doubling now. We are so far from catching up with ourselves that we have already entered a new phase of history while pretending that everything is the same way it has always been. Not so. But read on.

3
Why Now?

Why is all this happening now? Why not ten thousand years ago or ten thousand years into the future?

It is happening now because we have just completed a thirteen-thousand-year cycle and are reaching maximums on many different levels as a result. In order to explain this, I need to bring Drunvalo onto center stage.

Drunvalo says that there are two motions critical to our planet. The first, familiar for millennia, is the precession of the equinoxes (discussed below). The second is a wobble that was detected relatively recently. We, meaning the entire solar system, are spiraling through space in a manner that indicates we are attached to something. Astronomers noticing this began looking for that other body. It was first calculated down to a certain area of a particular constellation, then to a group of stars. In the early 1970s a specific star was targeted—Sirius A. We are moving through space with Sirius A in a spiral that is identical to the heliacal plane of the DNA molecule. We have a destiny with Sirius. As we move together a consciousness is unfolding much in the way the genes and chromosomes on the DNA molecule unreel their message from very specific places. There are key times when certain things can happen, when "genetically" critical alignments occur between Sirius and Earth and the rest of the cosmos. One very specific alignment is now happening.

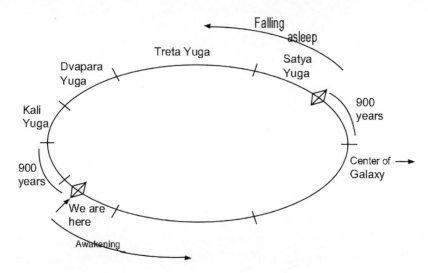

Figure 3.1. The precession of the equinoxes.

Now let's look at the precession of the equinoxes (fig. 3.1). The Earth's axis has 231/2 degrees of tilt. As the Earth goes around the sun, the axis remains tilted, giving us our four seasons.

But the axis itself is in a wobble, so the axis is changing as it goes around. It changes about 1 degree every 72 years. Every 2160 years this motion shifts our viewpoint of one constellation, and every 25,920 years it makes one full wobble. Science now however, is saying that a complete cycle takes 25,771 years; while the Mayans say it is 25,625 years.

Over slightly less than twenty-six thousand years then, the tip of the North Pole traces an ellipse. At one focus of that oval, it is closest to the center of the galaxy; at the other it is furthest away. What many ancients recognized, notably the Tibetans and Hindus, is that as we travel away from the center of the galaxy, we fall asleep, and as we turn the corner we begin to wake up.

These sages of old divided this ellipse into segments called *yugas*. Most of the current information on yugas was compiled in the last two thousand years, a period known as the Kali Yuga. This is coincidentally the most "asleep" point of the entire ellipse, so almost everything written about the cosmic cycle in the last two thousand years was interpreted by people asleep trying to extract from ancient writings they didn't even understand. Then they altered these writings in an attempt to conceal the truth. In other words, most of the information is not very trustworthy.

The ancients discovered two points located nine hundred years on either side of where we fall asleep and where we wake up. These points are each associated with tremendous change—changes of consciousness of beings and changes of poles on planets.

We sit right now at the turning point; in fact, December 22, 2012 was the first day of the new cycle. We are now moving back toward the center of the galaxy; we are waking up.

At 180 degrees opposite the last shift, the next one is imminent. That is why we are currently approaching our limits of population and environmental sustainability, among other things.

Binary Star System?

Walter Cruttenden wrote a book called *Lost Star of Myth and Time*. In it, he suggests that we are likely in a binary star system. He also backs it up with some real hard-core evidence.

One sign we would expect to see, according to Cruttenden, would be changes in the sun's rate of movement. In a binary system, orbital speed is not constant, and it could cause changes in the precession rate. When the two suns are closer to each other, the rate would increase, and it would decrease as they move farther apart.

In 2001, a team of scientists from the University of Michigan discovered that the asteroid belt at the edge of our solar system ends very abruptly. A sheer edge like they found would be unexpected and unexplained in a single sun system, yet, if our sun is indeed rotating around a companion star, a clear boundary to our solar system is both explained and expected.

All celestial bodies have angular momentum; a force that corresponds to their mass and motion. Yet, in our solar system, angular momentum is unevenly distributed. Our sun has 99.9 percent of the total mass, and only one percent of the total angular momentum. If we acknowledge that our sun is curving through space in a 25,920-year binary orbit, the sun's angular momentum was there all the time, primarily in its; orbital motion, and not just in its speed. This to me, is the most impressive and convincing of Cruttenden's arguments for a binary system

Virtually all, and perhaps all the star systems in the cosmos, are in a binary system so why would it be any different with respect to our sun? Though some refer to an unknown "Black Sun," perhaps the most likely candidate for this companion star is Alpha Centauri, our closest neighbor at 4.3 light years away.

According to Thoth, who ascended fifty-two thousand years ago, the degree of the Earth's pole shift at any time is directly related to consciousness on the planet and how much awareness will change. That is, there is a mathematical relationship between consciousness and degree of the pole shift. Another way to comprehend this is to realize that our cosmic falling asleep and waking up again are analogous to the day and night cycle of the Earth's twenty-four-hour rotation on its axis. For the most part, during the night the creatures of the hemisphere in darkness are asleep, and during the day most of them are awake. So it is with the cycle of the precession of the equinoxes. During the period of time in this cycle during which the planet is asleep, the male side takes over to protect us. It is always a female who leads us back into the light. According to Drunvalo, this has already happened. A woman took over sometime in February of 1989.

Pole shifts are a rather major planetary event; they are absolutely capable of sinking entire continents, as was the case thirteen thousand years ago. With that being said, it may well be a gentler ride this time around. Something happened in 1972 that changed everything. More later.

Where we are in the precession of the equinoxes tells us why everything is happening now. Sometime soon perhaps, we will most likely have experienced all these events. Most likely—but God can do anything. The other thing we must realize is that what is to happen will be determined by the thoughts and feelings of the people on the planet. If we change our consciousness, we can change the way in which the whole drama unfolds, regardless of any prophecy.

We are creating our reality. We are creating each new collective reality every moment. Our thoughts and feelings and actions are far more powerful than we could ever imagine. For the most part, we have thus far taken little or no responsibility for them. That however is changing so quickly that a whole new possibility never before dreamed of is emerging.

Pole Shifts

Pole shifts are connected, if Thoth is correct, with our evolutionary pattern. They are interrelated. The last pole shift was a major one and directly related to consciousness. It was not a positive shift; it was a negative one. We "fell" in consciousness.

Up until just recently, I think somewhere around 1950 or so, it was thought that polar shifts were something that never happened, or if they did it was once in a billion years and never again. But in fact, the poles shift every 12,500 to 13,000 years, or in other words, every time we come to these particular points in the precession of the equinoxes. So, it happens on a regular basis. Scientists are now finding that there indeed have been many pole shifts in the history of our planet.

Not only do pole shifts happen regularly, the poles have even reversed themselves, north to south and vice versa. Pole caps have been at the equator. There is hardly anywhere on the planet where you can't find seashells. You can find them on the top of the Rocky Mountains; you can even find them in Lake Titicaca, which was all below sea level at one time and is now more than twelve thousand feet high. Much of this information was discovered by scientists taking six-inch-diameter, eleven-foot-long core samples out of the ocean floor and reading the sediment just like tree rings.

Pole shifts are big. From methods of ionium and radiocarbon dating there is evidence to suggest that approximately thirteen thousand years ago the North Pole shifted from the Hudson Bay, or 60 degrees North Latitude and 83 degrees West Longitude, to its present location in the Arctic Ocean.[1]

Whenever there is a pole shift there is always a magnetic pole shift that happens just prior to it. Last time the magnetic poles went all the way to Hawaii, and then moved around and eventually came back to where they are now. The magnetic field of the Earth has been dropping in strength for the last two thousand years. This drop in strength began to accelerate five hundred years ago, then for the past forty years or so, the anomalies have become much greater. In the early 1990s the maps of airports had to be changed due to their sudden extreme inconsistency. When this value goes to zero, the poles shift.

Right now, we are physically on the line between the constellations Virgo and Leo. When we look out into the heavens, we see ourselves traveling from Pisces into Aquarius, but we are physically in Virgo going into Leo. This is why the Sphinx is a virgin and a lion together, the symbols for Virgo and Leo.

John White provides considerable data on this subject in his book *Pole Shift*. In northern Siberia, there is evidence of humans, mammoths, and trees, all shredded and ripped apart with incredible force and then instantly frozen solid. The mammoths had tropical food in their stomachs at the time. It should be noted that there are more than two hundred different kinds of ice formed as the temperature drops quickly. These relics are so deep down into the ice structure that this food is still edible today, 12,500 years later. Pole shifts happen very quickly. In fact, they complete themselves in about twenty hours.

There are all kinds of theories as to why they occur. A more modern theory, from the work of the Swedish physicist Hannes Alfvenis, is called magnetohydrodynamics or MHD. This theory proposes that beneath the Earth's solid crust there is a semisolid layer. This layer is solid as long as there is a magnetic field. When the magnetic field collapses, after fourteen days this layer becomes a liquid, allowing the Earth's crust to shift position.[2] A replica of this has been demonstrated in laboratories.

It has also been theorized that due to the massive buildup of ice in Antarctica, which is off center relative to the South Pole, that if the Earth's crust were suddenly free to move, that this ice mass would also move. This would force the crust to move to a new position. We would experience that as a pole shift.

Nobody knows for sure what triggers the process. Once it begins, the surface of the Earth moves at about two thousand miles per hour while the winds approach a thousand miles per hour. Obviously, that is enough to devastate just about anything. It's no wonder we hope this time to be able to do it with a little more control.

However large the pole shift may be, there will be a change in consciousness associated with it. That is, we will shift simultaneously. If we make a bigger consciousness change in advance, we can affect the physical pole shift even more. Our task, then, is a matter of becoming aware enough to control the way the next one happens so that it is enjoyable rather than fearful.

As a teacher of breathwork, let me say that a pole shift is just like birth. If a woman is in a state of fear, then giving birth will be difficult and painful. If she is relaxed and has no fear, then the birth happens easily. Birth can and should be easy. Either way, it is all a function of consciousness.

You might consider this information frightening. Don't be afraid. On a deeper level you already understand that there is no problem. Usually when a planet experiences a pole shift, the prognosis is that many are called and few are chosen. Eventually everyone makes it, though. A few people initially attain Christ-consciousness, and the rest drop down a dimension or two (which to them feels harmonious). Over a long period of time, probably hundreds of thousands of years, the few who made it bring the rest of the planet back and eventually the whole planet goes into Christ-consciousness. But here right now on Planet Earth something else is happening. I will lay this out in upcoming chapters.

4
Problems With Planet Earth

When it's time for consciousness to make a big shift, there is a simultaneous dying and a birth process leading up to the actual event. Planet Earth has major problems right now, much more severe than we are being told. If any one of these unresolved issues reaches its crisis point, all life on the planet will be severely threatened. And the problems are all coming to a head.

March 11, 2011 was the date of the massive 9.0-magnitude earthquake and subsequent tsunami waves that reached up to forty meters in height, breaching the Fukushima plant's sea defenses, triggering meltdowns of three of the six nuclear reactors.

I will have more to say about the Fukushima disaster, it is easily many times worse than the Chernobyl 1986 meltdown—but for now, let's consider the global impact of the resulting massive amounts of radioactive material that is being released into the Pacific. Our prevailing attitude seems to be that the oceans can handle whatever we dump into them. Do you really think the oceans can handle this?

Then there is the incredible accumulation of plastic concentrated in five offshore accumulation zones, the largest of which is known as the Great Pacific Garbage Patch (GPGP). It is located halfway between Hawaii and California. The GPGP covers an estimated surface area of 1.6 million square kilometers, an area twice the size of Texas.

A total of 1.8 trillion plastic pieces weighing about eighty thousand tons were estimated to be floating in the patch—a plastic count that is equivalent to two hundred fifty pieces of debris for every human in the world.

Plastic has increasingly become a ubiquitous substance in the ocean. Due to its size and color, sea animals confuse the plastic for food, causing malnutrition; it poses entanglement risks and threatens their overall behavior, health, and existence. Moving to the Gulf of Mexico, where on April 20, 2010, the Deepwater Horizon exploded, resulting in the deaths of eleven people and 210 million gallons of oil spewing into the water for eighty-seven days

A "blowout" on an oil rig occurs when some combination of pressurized natural gas, oil, mud, and water escapes from a well, shoots up the drill pipe to the surface, expands, and ignites. This disaster, easily the worst oil spill in U.S. history, was caused by a flawed well plan that did not include enough cement between the seven-inch production casing and the 9 7/8-inch protection casing. On July 15, 2010, British Petroleum (BP) announced that it had successfully plugged the oil leak using a tightly fitted cap.

For microscopic animals living in the Gulf, even worse than the toxic oil released during the disaster may be the very oil dispersants used to clean it up. More than 2 million gallons (7.5 million liters) of oil dispersants called Corexit were dumped into the Gulf of Mexico in an effort to prevent oil from reaching shore and to help it degrade more quickly. However, when oil and Corexit are combined, the mixture becomes up to 52 times more toxic than oil alone, according to a study published in the journal *Environmental Pollution*.

The spill has likely harmed or killed more than 80,000 birds, approximately 6000 sea turtles, and up to 25,000 marine mammals, including bottlenose dolphins, spinner dolphins, melon-headed whales, and sperm whales.

If we continue at our present rate, our oceans will die, and so will the plankton and phytoplankton. They are not only a cornerstone of the food chain but a major source of oxygen on the planet as well.

The Gulf Stream

The plot thickens. In an article published on February 22, 2004, by the *Observer/UK*, reporters Mark Townsend and Paul Harris told of a secret Pentagon report that warns of major European cities being sunk beneath rising seas as Britain is plunged into a Siberian climate sometime around 2020.

The report that was somehow obtained by the *Observer*, was commissioned by influential Pentagon defense advisor Andrew W. Marshall, director of the Pentagon's Office of Net Assessment. The actual report was authored by CIA consultant Peter Schwartz and Doug Randall of the California-based Global Business Network.

Then, in an article published in *Fortune* magazine on February 9, 2004, Marshall explained the following:

The Gulf Stream or scientifically referred to as the North Atlantic thermohaline conveyor is a stream of warm water that comes from south of the equator and flows over the surface of the ocean toward the north where this warm water keeps Northern America and Northern and Western Europe from freezing. It also holds most of the world's weather patterns in the way we are used to.

Then as this Gulf Stream cools down, it drops to the bottom of the ocean and returns as a river in the ocean to the south, where it warms up again and rises to the surface and then returns to the north one more time in a continuous convection current. It is a huge three-dimensional figure eight.

The motor that keeps this warm water flowing is found in the north where the Gulf Stream drops to the bottom of the ocean. It is the salt density of the ocean that causes this river to drop and pulls the warm water up from the south.

Now that the poles are melting and fresh water is flowing into the Atlantic Ocean and the salt density is decreasing, the Gulf Stream does not drop quite as far, which results in a slowing down of this

Stream. The Gulf Stream has been dramatically slowing down now for at least ten years.

As the Gulf Stream slows down, the warmth is not brought to the North Atlantic region, and the weather patterns begin to change for they are dependent on this warmth to keep a balance.[1]

The last time this happened was in 1300 CE, and it resulted in abrupt global climatic weather changes that lasted about 550 years. The east coast of America became extremely cold, while the Midwest and western areas grew hot and dry. Chaco Canyon in New Mexico did not receive rain for forty-seven years, causing the Anasazi Indians to leave.

The Gulf Stream actually stopped 8200 years ago, leaving Northern Europe under a half mile of ice, while New York and England endured weather similar to Siberia's. This lasted for a hundred years.

In a 2003 report entitled "An Abrupt Climate Change Scenario and Its Implications for United States National Security," authored by Peter Schwartz and Doug Randall, the Pentagon believes that is happening now, and the actual stoppage of the Gulf Stream would probably occur in three to five years from October 2003. This was their best guess and admittedly it was only a guess. This report intentionally considers the worst possible scenario, one that stretches the boundary of scientific plausibility.

However, if this does indeed begin to play out, it's time to head for the tropics folks! If the Gulf Stream were to stop completely, at a minimum, the Scandinavian countries would have to be completely evacuated, the growing of food would become very difficult if not impossible, and wars for food and water would break out all over the world.

As of this writing in March 2023, though none of these changes have yet taken place; we are told in a recent article appearing in the *Independent* (UK) that the Gulf Stream current, which has not been running at peak strength for centuries, is now at its weakest point in the past 1,600 years. They go on to warn us that catastrophic changes in global weather patterns could be on the horizon, noting that there could be weather disruption across the U.S., Europe, and Africa, and sea levels could rise rapidly on the east coast of the United States.

And then of course Hollywood had to chime in with their two-cents worth with the May 2004 release of *The Day after Tomorrow,* a 20th Century Fox blockbuster disaster movie with a similar premise. This movie didn't just stretch the boundary of scientific plausibility, it took it to new levels of absurdity.

Fukushima

Now let's get back to Fukushima. First let me say that nuclear fission is against galactic law, and we are now beginning to understand why. Fission, splitting matter apart, is quite different from fusion. All the suns are fusion reactions and that's okay.

It's not easy to gather factual information on Fukushima, considering that the nuclear power agency among others, has a vested interest in protecting the illusion of safety and in hiding the truth about the inherent dangers of nuclear reactor meltdowns. The Japanese government in 2013 passed a secrecy act which mandates up to ten years in prison for civil servants who "leak secrets," and as much as five years for journalists who are brave enough to leak concealed information. Powerful vested interests want to bury the truth.

One of the best sources I've found was a November 22, 2017, article written by Robert Hunziker on Defend Democracy Press. The following is from that article:

According to Mr. Okamura, a TEPCO manager, the current status as of November 2017 was as follows: "We're struggling with four problems: (1) reducing the radiation at the site (2) stopping the influx of groundwater (3) retrieving the spent fuel rods and (4) removing the molten nuclear fuel." (Source: Martin Fritz, The Illusion of Normality at Fukushima, Deutsche Welle–Asia, Nov. 3, 2017.)

In short, nothing much has changed in nearly seven years at the plant facilities, even though tens of thousands of workers have combed the Fukushima countryside, washing down structures, removing top-soil and storing it in large black plastic bags, which end-to-end would extend from Tokyo to Denver and back.

Complete nuclear meltdowns are nearly impossible to fix because, in part, nobody knows what to do next. That's why Chernobyl sealed off the greater area surrounding its meltdown of 1986. Along those same lines, according to Fukushima Daiichi plant manager Shunji Uchida:

"Robots and cameras have already provided us with valuable pictures. But it is still unclear what is really going on inside."

Seven years and they do not know what's going on inside. Is it the China Syndrome dilemma of molten hot radioactive corium burrowing into Earth? Is it contaminating aquifers? Nobody knows; there is no playbook for one hundred percent meltdowns. Fukushima Daiichi proves the point.

"When a major radiological disaster happens and impacts vast tracts of land, it cannot be 'cleaned up' or 'fixed.'" (Source: Hanis Maketab, Environmental Impacts of Fukushima Nuclear Disaster Will Last 'decades to centuries'—Greenpeace, Asia Correspondent, March 4, 2016.)

Meanwhile, the world nuclear industry has ambitious growth plans, 50-60 reactors currently under construction, mostly in Asia, with up to four hundred more on drawing boards. Nuke advocates claim Fukushima is well along in the cleanup phase so not to worry as the Olympics are coming in a couple of years, including events held smack dab in the heart of Fukushima, where the agricultural economy will provide fresh foodstuff.[2]

Then heaven forbid, what if there is another major earthquake—after all, Japan is earthquake country.

According to Dr. Shuzo Takemoto, professor, Department of Geophysics, Graduate School of Science, Kyoto University:

"The problem of Unit 2. . . . If it should encounter a big earth tremor, it will be destroyed and scatter the remaining nuclear fuel and its debris, making the Tokyo metropolitan area uninhabitable."

(Shuzo Takemoto, Potential Global Catastrophe of the Reactor No. 2 at Fukushima Daiichi, February 11, 2017).

Meanwhile back at Chernobyl, thirty years after its explosions and fire, in November 2016, a huge metal cover was moved into place over the wreckage of the reactor and its crumbling, hastily erected cement tomb. The giant new cover is 350 feet high, and engineers say it should last 100 years—far short of the 250,000-year radiation hazard underneath.

Solar Flares

Okay, as massive as this still totally out of control disaster at Fukushima is, it is but a Sunday stroll in the park relative to the ongoing threat of a massive solar flare. Considering that such an event could easily wipe out the power grid—leaving us with no power, no transportation systems, no communication systems, no banking, no internet, no food, and no water delivery systems. This would truly be an end of the world as we know it situation.

Now, to add to the drama, consider the fact that this already has happened, in 1859. It was known as the "Carrington event," where the technology of the day, the telegraph systems worldwide, were knocked out.

Now let's consider the impact such an event would have on our 440 nuclear power plants operating across thirty countries around the world today. According to a chilling yet excellent article posted by Mike Adams of *Natural News*:

All nuclear power plants are operated in a near-meltdown status. They operate at very high heat, relying on nuclear fission to boil water that produces steam to drive the turbines that generate electricity. Critically, the nuclear fuel is prevented from melting down through the steady circulation of coolants which are pushed through the cooling system using very high-powered electric pumps.

If you stop the electric pumps, the coolant stops flowing and the fuel rods go critical (and then melt down). This is what happened in

Fukushima, where the melted fuel rods dropped through the concrete floor of the containment vessels, unleashing enormous quantities of ionizing radiation into the surrounding environment. The full extent of the Fukushima contamination is not even known yet, as the facility is still emitting radiation.

It's crucial to understand that nuclear coolant pumps are usually driven by power from the electrical grid. They are not normally driven by power generated locally from the nuclear power plant itself. Instead, they're connected to the grid. In other words, even though nuclear power plants are generating megawatts of electricity for the grid, they are also dependent on the grid to run their own coolant pumps. If the grid goes down, the coolant pumps go down, too, which is why they are quickly switched to emergency backup power—either generators or batteries.

As we learned with Fukushima, the on-site batteries can only drive the coolant pumps for around eight hours. After that, the nuclear facility is dependent on diesel generators (or sometimes propane) to run the pumps that circulate the coolant which prevents the whole site from going Chernobyl. And yet, critically, this depends on something rather obvious: The *delivery of diesel fuel* to the site. If diesel cannot be delivered, the generators can't be fired up and the coolant can't be circulated. When you grasp the importance of this supply line dependency, you will instantly understand why a single solar flare could unleash a nuclear holocaust across the planet.

When the generators fail and the coolant pumps stop pumping, nuclear fuel rods begin to melt through their containment rods, unleashing ungodly amounts of life-destroying radiation directly into the atmosphere. This is precisely why Japanese engineers worked so hard to reconnect the local power grid to the Fukushima facility after the tidal wave—they needed to bring power back to the generators to run the pumps that circulate the coolant. This effort failed, of course, which is why Fukushima became such a nuclear disaster and released countless becquerels of radiation into the environment (with no end in sight).

And yet, despite the destruction we've already seen with Fukushima, U.S. nuclear power plants are nowhere near being prepared to handle sustained power grid failures.[3]

What's the chance of all this actually happening? A report by the Oak Ridge National Laboratory said that "over the standard 40-year license term of nuclear power plants, solar flare activity enables a 33 percent chance of long-term power loss, a risk that significantly outweighs that of major earthquakes and tsunamis."[4]

These are just a few of the many environmental problems we face. There are so many going on right now that we seem to occupy a sinking boat. A particular aspect of planet Earth, this third-dimensional aspect which we call our daily lives, will not be intact much longer.

If you were to return here in just a few years you would find the planet dead on a third-dimensional level. But life would still exist on Earth; we are all going to shift up in wavelength to another place that is prepared for us, which is beautiful and where there are no problems. We will move into a slightly shorter wavelength and a higher energy vibration. The Bible refers to this state as Heaven. Actually, we will be going from the third to the fourth dimension.

According to Drunvalo, extraterrestrials regularly visit a planet like ours. But it is against universal law for them to interfere with us. For this reason, they enter one overtone higher than the one the planet sits on so that they are invisible to us. But they can monitor us very clearly from this higher vibratory rate. In fact, the next higher overtone on our planet is right now so full of vehicles filled with curious occupants that more recently arriving visitors have had to go into the second overtone, and they have now almost completely filled that one. Ours is the most unusual situation ever known in this universe. There are even beings from distant galaxies here to watch. Usually, they would never bother with us. Most of the ETs who are here are not only in light bodies but they are of such a nature that their ship and their body are one and the same.

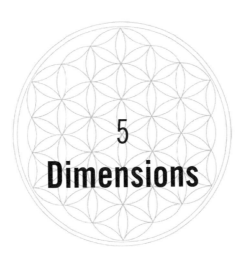

5
Dimensions

When I talk about dimensions, what do I mean? If I was a mathematician, I'd probably be telling you about things like hyperspace or maybe 4D or 6D^3. But I don't mean any of that. *Star Trek* came pretty close in a few of their episodes—one in particular called "Mirror, Mirror," comes to mind, where Kirk, Scotty, McCoy, and Uhura due to a transporter malfunction, accidentally switch places with their malevolent counterparts from a parallel universe. Of course the wrong guys were beamed up to the Enterprise too. When will they ever learn? Using the transporter during an ion storm can be a very iffy proposition indeed!

Here's what I really mean. Every universe that has ever been created is right here, passing through the room in which you are sitting, and they are all interlinked. The only difference between one and the next is wavelength. People in physics almost all agree: everything is sine wave, or wavelength, and wavelength is the key to everything (fig. 5.1).

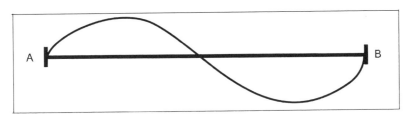

Figure 5.1. The distance from A to B, measured in a straight line, is wavelength.

With wavelength, you have a wide range of possibilities; you have some that are unbelievably tiny, and some that are so long that they would look almost like straight lines. And when you put them all together, you have the electromagnetic spectrum.

Our reality, ranging from the most distant galaxy to subatomic particles, and including the space in which we live, is one universe, and it has a wavelength of 7.23 centimeters.

This particular wavelength was discovered quite by accident, when two astronomers, Robert Wilson and Arno Penzias—working for Bell Telephone Laboratories—pointed their fifty-foot antennae skyward, and discovered a low-level hiss, a constant noise that seemed to be coming from everywhere. This was in 1964 when they were installing their microwave system; they were using a wavelength of around 7 centimeters. It was Drunvalo who suggested the specific wavelength of our third-dimensional reality is 7.23 centimeters.

In quantum physics, every object can be looked upon either as particles (atoms) or vibration (wavelength), and every particle or piece of matter has its own sine wave signature. If you took all the objects in our universe and averaged them by wavelength, you would find this average to be 7.23 centimeters.

Dimensions are separated from one another by wavelength in exactly the same way that notes are separated on a musical scale (fig. 5.2). Each tone on the scale sounds different because of its wavelength. Any octave on the piano has eight white keys and five black keys, which together give its player the chromatic scale. The thirteenth note is actually the first note of the next octave, and these octaves keep repeating themselves in either direction.

Music, as is the electromagnetic spectrum, is holographic. That means then, that in between any two notes, you've got a smaller version of the entire scale. And the same is true with regard to the different dimensional worlds. So, you've got the twelve major dimensions, and in between each dimension, you have a smaller version of the whole; these

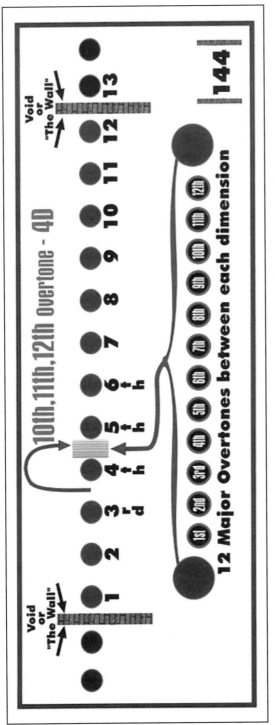

Figure 5.2. Dimensions are separated by wavelength in exactly the same way that notes are separated on a musical scale. There are a hundred and forty-four dimensions in each octave. We on Earth are in the third dimension and beaded for the tenth, eleventh, or twelfth overtone of the fourth dimension.

are called overtones. The thing about these overtones though, is that every single one of them is a universe just as vast as this one. Twelve times twelve gives you one hundred and forty-four different dimensional worlds.

Now just as you have on a piano, there is more than just one octave; they just keep repeating themselves in both directions. The thirteenth dimension happens to be the first dimension in the next octave; and they keep repeating themselves a number of times in both directions. And while it's not an infinite number of dimensions; it hardly matters—you'd probably have to live a few billion years in order to have enough time to count them all.

To keep it simple, it's really similar in a lot of ways to a television set. When you turn on your set, what you're doing is tuning to a specific channel—which is really a certain wavelength transmission. When you change the channel, you're tuning to a different wavelength and a different image appears on your screen.

There is a void between dimensions like the void between two notes. However, the voidness between dimensions is a total nothingness—no directions, no light, and no nothing—just a total sensory-deprivation experience. There is a greater void—a great wall of voidness if you will, between octaves. Each dimension is also separated from the others by a ninety-degree rotation. If you could change wavelengths and rotate ninety degrees, you would disappear from this world and reappear in whatever dimension you were tuned to—and you'd be just as real there as you are here. The images you see looking out from your eyes would change according to the wavelength of the world you had entered.

This planet has many different levels; the ascended masters for example, are residing on a much higher level of the Earth's consciousness. They are primarily on the tenth, eleventh, and twelfth overtones of the sixth dimension. So, there's a whole lot more to planet Earth, as well as all other planets and suns; there's other worlds, and they're all right here. But our consciousness is tuned to one particular wavelength, the third dimension—the place that we call home. Meanwhile, we literally exist on multiple dimensional levels, and our experience on each level is completely different.

For example, if we were to go up one level, which we are in the process of doing, we would find that whatever we think, as soon as we think it, instantly manifests. Here, by contrast, on the third dimension there is a time delay. Even though our thoughts create our reality unerringly here as well, their manifestation obviously is not instant.

The key to understanding how to move from one dimensional level to another begins with locating an electromagnetic field shaped in the form of a star tetrahedron. That field is around our bodies.

The Merkaba

A star tetrahedron is made up of two interlocking tetrahedrons in a manner that resembles the Star of David but three-dimensionally (fig. 5.3). The two interlocking tetrahedrons represent male and female

Figure 5.3.
A star tetrahedron.

energy in perfect balance. The tetrahedron facing up is male, and the one facing down is female. There is a star tetrahedral field around everything, not just our bodies.

There is also a tube that runs through the body. It connects the apexes of this star tetrahedral field. Learning how to breathe through this tube, combined with rotating the fields, produces the *Merkaba* (sometimes spelled Merkavah and, or Merkabah), a vehicle of ascension.

We have a physical body, a mental body, and an emotional body, and they all have star tetrahedral shapes. These are three identical fields superimposed over each other, the only difference among them being that the physical body alone is locked—it does not rotate. The Merkaba is created by counter-rotating fields of energy. The mental star tetrahedral field is electrical in nature, male, and rotates to the left. The emotional star tetrahedral field is magnetic in nature, female, and rotates to the right. It is the linking together of the mind, heart, and physical body in a specific geometrical ratio and at a critical speed that produces the Merkaba.

The word "Mer" denotes counter-rotating fields of light, "Ka" spirit, and "Ba" body, or reality. So the Mer-Ka-Ba then is a counter-rotating field of light that encompasses both spirit and body, and it's a vehicle—a time-space vehicle. It's far more than just that, in fact there isn't anything that it isn't. It is the image through which all things were created, and that image is around your body in a geometrical set of patterns. That image begins at the base of your spine as small as the original eight cells from which our physical bodies first formed. It moves out from there, first forming a star tetrahedron, then a cube out at your hands, then a sphere, then back-to-back pyramids (known as an octahedron,) then an icosahedron, and then a dodecahedron. (You will become familiar with these shapes as this information is unfolded.) The field extends out a full fifty to sixty feet in diameter, depending on your height.

That field is an immense science that is being studied everywhere throughout the cosmos. How well someone understands the Merkaba, is usually in direct relationship to their consciousness level.

Again, the counter-rotating fields of light of the Merkaba comprise a time-space vehicle. Once you know how to activate these fields, you can use your Merkaba to travel throughout the universe.

On most terrestrial people the Merkaba is not functioning; there are only a few thousand people on this planet whose Merkabas are functioning, and about eight thousand ascended masters who reside on another level of the Earth's consciousness.

6
Earth History

In mathematics there are many different numerical configurations known as mathematical sequences. Some examples: 1,2,3,4,5,6,7,8,9; 1,1,2,3,5,8,13,21,34,55; and 2,4,8,16,32,64,128,256,512. In every case you need only three consecutive numbers to figure out the entire sequence. For example, if you have 4,5,6 or 5,8,13 or 2,4,8 you can determine the entire sequence.

Our minds are arranged in three components. In polarity consciousness we think of duality, but in fact there is no such thing. For every polarity you can think of, there will always be that third component. For example, for up and down, there is middle; for black and white, there is gray; for hot and cold, there is warm. Space is arranged in the same way: there is macrocosm, the space in which we live, and microcosm. Time too: past, present, and future.

In order for us to understand what is happening now and what will happen in the future, it is essential that we know the past. We have to understand what happened before, combined with what is now so we can know what will happen. Plants, for instance, use a mathematics described by the Fibonacci sequence to determine what to do in their growth. A plant will grow one leaf, and then it will grow one more. Before it continues, it looks back to see what it just did, ascertains where it is now, and then knows where it is going. It adds the number of leaves it just grew to its present complement to know how many leaves to grow next. It says, for example, I just grew one leaf and I am at one, so that means I grow two. When it gets to two leaves it says, I was just at one and I grew two more, so now I am at three, and so on. It needs to look back to see what it just did combined with where it is now to know where to go. Such is organic structure.

As humans we have to know our history because we have to know how we got into our present predicament in order to get out of it. We are told that it all began in 3800 BCE in Sumeria, and that nothing came before that except hairy barbarians. We assume we are the greatest thing that ever lived on this planet. In fact, there have been civilizations so far ahead of us that we can't even imagine them.

For all you Trekkies, you may recall from the original series, an episode called "Errand of Mercy". From imdb.com we find the following storyline:

> War! The Klingons and the Federation are poised on the brink, and then war is declared. Kirk and Spock visit the planet Organia. Organia, inhabited by simple pastoral folk, lies on a tactical corridor likely to be important in the coming conflict. Whichever side controls the planet has a significant advantage. But the Organians are a perplexing people, apparently unconcerned by the threat of the Klingon occupation or even the deaths of others in their community. Finally, Kirk and the Klingon commander Kor learn why, and the reason will change Federation/Klingon relations for decades to come.

Ayelborne, the head of the local council, along with fellow council members reveal their true nature beings of pure energy as they all disappear into a ball of light. Both sides are instantly incapacitated, forcing them to agree to a cessation of hostilities.

Mr. Spock summed it all up quite succinctly with the following remark, "I should say the Organians are as far above us on the evolutionary scale, as we are above the amoeba."

Civilizations on this Earth go back five hundred million years. The planet is a star seed to which external life forms have come from all over, combined with each other, generated new life forms, and left. Each of the life forms they created has gone through five levels of consciousness. Right now, we are on the second of the five.

Almost all the evidence from past advanced civilizations has been put out of context or ignored. For instance, we have a connection with the star Sirius about which little is known but which is essential to understanding our present plight. Robert Temple's book *The Sirius Mystery* presents the following: There is a tribe near Timbuktu in Africa called the Dogon. For more than seven hundred years this group has had information that presumably it cannot have—information that our scientists have only known recently.

The Dogon know about the star Sirius in detail. Sirius is the brightest star in our sky, situated to the left and straight down from the belt of Orion. The Dogon told researchers decades ago that there is another tiny star rotating around Sirius and made of the heaviest matter in the universe. This star was said to complete its rotation once every fifty years. It is a very old star. We didn't begin to know of this until 1844 when a German astronomer named Friedrich Bessel first discovered it. We first saw it through our telescopes in 1862. Like the one in Dogon "myth," it is very old. The orbit of the star was calculated to be 50.1 years. This star was named Sirius B, and the original Sirius renamed Sirius A.

When a team of scientists visited the Dogon tribe to determine how they knew about this, the elders said that a flying saucer landed. Beings emerged and made a large hole in the ground which they filled with water. The occupants, who looked like dolphins, jumped into the lake they had made then came up to shore and talked to the Dogon people, telling them that they were from Sirius and relating many Sirian stories.

The Dogon had an even more incredible bit of information. They had a visual image of the movement of Sirius A and Sirius B from Earth for the time period between 1912 and 1990, culminating in an exact image of where these stars would be at this time. They also had a great deal of information about the planets in our solar system, including various moons. How did they know or imagine this so specifically?

The Uros Indians, who predate the Incas and who live on artificial floating islands in Lake Titicaca, have a similar story. It all began, they say, when a flying saucer came out of the sky at Lake Titicaca and landed on the Island of the Sun. And according to them, there was a great deluge that left the entire Earth submerged. There was a sacred rock on the Island of the Sun, which was the first tip that broke out of the water. And as we shall soon see, it was there that a man named Chiquetet Arlich Vomalites along with several Atlantians, landed and began the Inca race. Their original story tells of earlier encounters with Dolphin-like beings who came in a flying saucer, landed on the Island of the Sun and began to communicate with the locals.

My next bit of information has to do with the Giza Plateau. The Sphinx and the Great Pyramid are two of the greatest enigmas of all time. The exact dates of their construction are still cause for debate. How did ancient people develop the technology to build the Great Pyramid with a skill and accuracy still unmatched today?

The Great Pyramid stands more than four hundred fifty feet tall; it weighs more than six million tons. It has a footprint in excess of thirteen acres. It's perfectly aligned to true north, south, east, and west. To achieve that precision of alignment with a monument of this scale is an amazing technological feat, and one that is continually overlooked by our scholars today.

This extraordinary monument, a high-tech achievement by any standards, appeared on the desert. This is a great mystery, totally unexplained by conventional history. And what does it say about the origins of human civilizations and our view of the past and where we came from?

Could it be that we've been missing a huge part of the human story—like we've developed a rather severe case of dementia, and we've somehow lost the record of our true history, going back many thousands of years? What if we could go back into that dark epoch, what would we discover about ourselves?

Egypt is not the only region where ancient ruins proactively taunt the mind with the mystery of how they came into being. There are many such places, such as Baalbek in Lebanon where massive eight-hundred-ton blocks were moved at least one-third of a mile and then positioned high atop a wall.

The entire ancient complex of Baalbek was constructed with huge megalithic blocks that range in weight from one hundred to fifteen hundred tons each.[1] How is it possible that ancient people managed to transport these huge blocks of stone and then place them in position so that they fit perfectly with each other? Mainstream archaeology has no clue!

Better yet, the precision found at Baalbek is fascinating, resembling other ancient sites in South America. The precision of the megaliths is breathtaking; they were arranged in such a way that you cannot fit a single sheet of paper in-between them.

Then there's the great ceremonial center known as Tiahuanaco in Bolivia. Not as old or as imposing as the pyramids in Egypt, what makes this site so unusual is its unique location. Many believe that it could not have been built without intervention from an advanced life form.

The biggest blocks at Giza weigh two hundred tons, the biggest blocks at Tiahuanaco weigh four hundred tons. That's four-hundred-ton blocks of stone used to create enormous construction at twelve thousand feet above sea level, in an area—because of the altitude—where it's almost impossible to grow food. And yet we're asked to believe that this was done with massive labor forces hauling these blocks along on ropes. Sure, I believe that, don't you?

So, we're looking at the evidence of a technology that we don't understand, that like the Giza monuments, also incorporates extremely precise astrological alignments.

Flash, this just in, we now know how they did it, the Egyptians that is. They wet the sand first! Remember now, the next time you want to move two-hundred-ton blocks around the desert, just wet the sand and it's a piece of cake! Of course, that doesn't explain how they raised them up to four hundred and fifty feet in the air. Oh well!

And what about the Nazca lines in Peru? Covering more than two hundred square miles, a bewildering pattern of gigantic artwork litters the Nazca plateau. In addition to figures of birds, spiders and animals, straight lines stretch out in all directions. Believed to have been laid down here more than two thousand years ago, what was their purpose? Observable only from the air, is this a canvass of signs and symbols etched in the sand to honor pagan gods, or is it a set of messages aimed at travelers from the stars?

Meanwhile, back at the Sphinx, Egyptologists say that the Sphinx was built around 2500 BCE by the pharaoh Chephren. In his 1961 book called *Le roi de la theocratie pharaonique* (*Sacred Science*), the mathematician, philosopher, and Orientalist R.A. Schwaller de Lubicz notes:

> We have to acknowledge that a great civilization must have preceded
> the vast movements of water that passed over Egypt; it is this which
> is implied by the existence of the sphinx sculpted in the rock on the
> western cliffs at Giza, this sphinx whose whole leonine body, with
> the exception of the head, shows an indisputable water erosion.[2]

An Egyptologist named John Anthony West, after reading de Lubicz' book in 1972, decided to research the weather patterns found on the Sphinx. He saw that the wear patterns were excessive, up to twelve feet deep in the back. He brought in an American geologist, Robert Schoch, to look at the Sphinx from a geological point of view.

He discovered beyond any doubt that the patterns were not the result of wind and sand but water. It was calculated that there would have to have been a minimum of a thousand years of torrential rain flowing on the Sphinx consistently for it to have been worn to these patterns.[3] Geology now stands in direct contradiction with archaeology. The Sahara Desert is at least seven to nine thousand years old, which means the Sphinx has to be eight to ten thousand years old, minimum.

Egyptian archaeologists have nothing to say about this. The evidence is overpowering, but the truth threatens to break down all our concepts of who was on this planet when we believe that there was no one here capable of building something like the Sphinx (eight to ten thousand years ago). According to Thoth, the Sphinx contains proof of five and a half million years of civilizations on this planet, even though there have been more like five hundred million years of civilization. Something happened five and a half million years ago that broke the Akashic Record memory of the Earth. Even Thoth doesn't know what it was and or how to get through to the older records.

Archeologists have often pondered on the exact process which led to civilization as we know it. For it was with the suddenness of a sunrise that human ingenuity and engineering skills arose from the Stone Age and burst forth onto the landscape of history. The Sumerian civilization seemed to arise overnight with no evolution at all. The same is true for Egypt. Egyptian writing appeared one day in its most developed form and went downhill after that. All the ancient civilizations—Sumer, Babylonia, Egypt, and so on—developed quickly and went downhill immediately.

In the ruins of Sumerian cities archaeologists have found tablets that depict the solar system, listing the planets in their correct order. One even gives the distances between the planets. How did anyone know this? The ancient tablets also provide detailed records concerning the precession of the equinoxes.[4] It has been calculated that the only way to discover the precession of the equinoxes is from observation, and that the minimum time of watching the night sky would have to have been 2,160 years. How did the Sumerians have this information, when, according to our way of thinking, there was no advanced civilization 2,160 years before them?

Perhaps this was all the work of hairy barbarians. If so, those guys sure had more on the ball than we've given them credit for!

Our Creators

Now my discussion will drop back about four hundred thousand years in our history. Here I will incorporate information from both Thoth and researcher and author Zecharia Sitchin, in particular Sitchin's books *The 12th Planet* and *Genesis Revisited*. Sitchin believes that there is another planet in our solar system, the Sumerian Nibiru, which has an elliptical orbit similar to that of a comet. It takes thirty-six hundred years for the planet to make one complete circle of the sun. The people on this planet, called the Nefilim, came to Earth more than four hundred thousand years ago. Thoth doesn't say why; in fact, he never mentioned the Nefilim by name. He only told Drunvalo that there were giants on the Earth at the time (the Nefilim ranged from ten to sixteen feet in height). Sitchin claims the reason they came was that they needed gold for their atmosphere. In *Genesis Revisited*, he writes:

> On their planet Nibiru, the Anunnaki/Nefilim were facing a situation we on Earth may also soon face: ecological deterioration was making life increasingly impossible. There was a need to protect their dwindling atmosphere, and the only solution seemed to be to suspend gold particles above it, as a shield.[5]

So, they came here to mine gold. After 200,000 years or so of enforced labor, the miners rebelled and decided to create a subservient race—which is us—to mine the gold for them. It is notable that in southern Africa in the oldest known gold mines, archaeologists have found the bones of *Homo sapiens* and artifacts that go back at least

50,000 to a 100,000 years. Sitchin asserts that the Nefilim created us about 300,000 years ago, but Thoth is very exact. Thoth says we were created exactly 200,239 years ago (from 2023).

Sitchin theorizes that the Nefilim created us through genetic experiments, but according to Thoth they couldn't do it alone. They had to have help from outside the solar system. That external help came from a familiar source. The Nefilim first landed in the ocean and emerged as half-men and half-fish. They went underwater initially to make contact with the dolphins and the whales, who were and remain to this day, the highest levels of consciousness on the planet. They had to check in and get permission to do what they wanted to do. The dolphins and whales came here from the third planet out from Sirius B. There is a small humanoid population on this mostly water planet, but the majority of the inhabitants are cetacean. As we shall soon see, the Sirians played a co-creative role in our creation process. So, in essence, the Nefilim when they first landed, were getting permission from the Sirians. According to the Sumerian records, the Nefilim first went to southern Iraq and built their cities. They then went to southern Africa to mine gold.

The Sumerian records are round clay tablets that came out of the ancient Sumerian cities, the cities mentioned in the Bible. These records, discovered only in the last century, are the oldest sources of information, bar none, on the planet. Sitchin is one person who has the ability to interpret these records.

Again, Thoth said there were giants and that they became our mother. He said that seven of these beings dropped their bodies and formed spheres of consciousness. They merged into the seed of life and created an ovum. When seven beings link together geometrically in this way to form the seed of life, a flame appears, four feet tall and of bluish-white light. It is cool but looks like a flame. This was then set in the "Halls of Amenti."

The Halls of Amenti is a very ancient place, built more than five and a half million years ago. No one knows how old the Halls of Amenti really are or who erected them, because of that event five and a half million years ago that broke the Akashic records of the planet. Remember—even though history on the planet goes back five hundred million years, we have access only to the last five and a half million years.

The Halls of Amenti is actually a dimensional warp in space resembling a womb. There is only one way in, but once you get there it is like being in infinite space. Such a warp sits always one-dimensional overtone higher than the vibrational level of the Earth. It is located usually 440,000 miles out in space, but during the era of Atlantis it was on the surface of the Earth. Now, it is 1000 miles inside the Earth.

Simultaneous preparation for this creation or trans-semination was being made on Sirius B or, more precisely, the third planet out from Sirius B. Sixteen males and sixteen females who comprised a married family there traveled to Earth from Sirius B and went directly to the flame in the Halls of Amenti. They lay down and merged with the flame. Their conception period here was two thousand years. These two separate races were involved in our creation—one from Nibiru and one from Sirius.

Thoth said that we were originally placed on an island off the coast of southwestern Africa called Gondwanaland, and that we were placed there primarily so we could not leave. We were there fifty thousand to seventy thousand years. From there we were brought to Southern Africa. Interestingly enough, African creation stories all agree on one piece of information. They all say their people came from an island off the southwestern coast of Africa, called Gondwanaland.

According to Sitchin's interpretation of the Sumerian texts in *The 12th Planet*, after the Nefilim created us to work in the gold mines of Africa, some of us were brought to Mesopotamia to help in the gardens in E.DIN. The "gods" loved us; because, after all, we were made in their own image. But in the garden in E.DIN where the Nefilim had their orchards, we were told not to eat the fruit of a certain tree called the "tree of knowledge of good and evil." The records specifically mention Adam and Eve by name, and they disobeyed. Eating this fruit and gaining its "knowledge" was significant because in addition to giving us polarity consciousness, it also gave us the ability to reproduce sexually. Up until this point, we were hybrids, a cross between two different species, and we were incapable of reproduction. Sitchin interprets the Sumerian texts as saying that modern humans are a cross between the Nefilim and *Homo erectus,* the predecessor of *Homo sapiens.* It is through Thoth that we get the additional information about the role of the beings from Sirius.

Not surprisingly, the Nefilim did not want us reproducing. They wanted to maintain control of their own experiment. The knowledge we gained from eating the fruit was not scientific as such; it was the knowledge of how to procreate, how to turn ourselves from sterile hybrids into a new species fully capable of reproducing. The Nefilim were angry when we gained the ability to reproduce; and they especially didn't want us eating from the other tree—the tree of life, for that would have made us immortal—so they made us leave the garden. According to scholars of ancient texts, the Sumerian records precede the biblical records, and the biblical creation stories seem to be a summary of older Sumerian texts.[6]

Although we had to leave their garden, the Nefilim allowed us to grow food on our own. We went to the mountainous area east of the gardens in Mesopotamia. So at that point, there were two versions of humans; those who could procreate and were free, and those who were hybrids and were gold-mining slaves. According to Thoth, it remained that way for a long time. But then there was a big shift in consciousness and another pole shift, and Gondwanaland sank. Many of the survivors went to Africa, but the most evolved—those of the Adam and Eve lineage—went to Lemuria, a land that rose above water when the other continent sank.

Lemuria

The continent of Lemuria was really a series of islands that extended from the Hawaiian Islands down to Easter Island. It lasted for sixty thousand to seventy thousand years. During the era of Lemuria—

where we were allowed to freely develop on our own—life was good, we were learning much and we were evolving quickly. The consciousness of the Lemurians became predominantly intuitive and female, and by the time that the continent sank, we were about the equivalent of a twelve-year-old female. The Lemurians had technology that we can't even begin to understand—for example, dowsing rods that work only when your mind and heart link together.

Moving back in time to Lemuria approximately eighty thousand years ago, or about a thousand years before the continent sank—and the continent was sinking very slowly almost from the beginning—there lived a couple by the name of Ay and Tiya. They had become physically immortal beings during the course of their lives and so they opened a school to teach immortality and ascension to others—the Naacal Mystery School. Ascension is a method of consciously moving from one world to another, taking your body with you. It is different from resurrection, which is consciously moving from one world to another by dying and then reforming your light body on the other side. In the course of its existence, the school graduated about a thousand immortal masters right up until the time when Lemuria was sinking rapidly. Extremely intuitive, the inhabitants knew their land was submerging; they were well prepared for it, so there were probably very few casualties. As the continent became uninhabitable, almost everyone from Lemuria migrated to a zone as far south as Lake Titicaca and as far north as Mount Shasta.

Atlantis

As Lemuria sank, the poles shifted, and the land mass of Atlantis arose. The thousand or so immortal masters of the Naacal Mystery School of Lemuria went to Atlantis, specifically to one of its ten islands to the north called Unal. When they arrived, the first thing they did was build a wall right down the middle of the island from north to south.

This wall, which was about forty feet high and twenty feet wide, sealed off both sides so you could not cross from one to the other.

Next, these beings of light erected a smaller wall from east to west, thereby dividing the island into four quadrants. This structure replicates the human brain, which is divided into two hemispheres with the corpus callosum running down the middle. The left hemisphere, the male side, is based on logic; the right hemisphere, the female side, is based on experience or intuition. But the male side also has a female or experiential aspect associated with it, and the female side has a male or logical component. These are the four quadrants of the human brain.

When they had completed the division of their island, half of the masters went to one side and the other half went to the other side. The masters on the left side became the logical or male component, while the ones on the right side became the intuitive or female part of the brain. They carried this out to the point that after a few thousand years, it came time to choose three people to be the corpus callosum to link to left and right hemispheres. In so doing, the "brain of Atlantis" became alive.

And then this "brain" projected onto the main island the ten patterns of the Tree of Life, so that vortexes of energy began to rotate out of these ten spots and summon the Lemurians to Atlantis. Each person was drawn to the specific vortex that was associated with his or her true nature. The Lemurians who had settled from Lake Titicaca to Mount Shasta didn't know why, but they suddenly felt the need to migrate to Atlantis. They were drawn there by the energy vortex created by the immortal masters.

Unfortunately, the evolutionary pattern of Lemuria was such that they had only developed the nature of eight of the ten vortexes associated with the Tree of Life. The Lemurians migrated to eight of these ten spots, which became major cities, but the remaining two vortexes were left vacant. This is where a big problem first began.

The vacant energy vortexes ended up pulling in two extraterrestrial races that then joined with our human consciousness and became part of our evolutionary pattern. The first race was the Hebrews, whose origin is unknown. Thoth said that they were like a child who had failed the grade and had to step back and repeat it. They came with full permission from Galactic Command and were not a problem. In fact, in many ways they helped because they brought in advanced information that we didn't yet have. The problem was with the second race. Specifically, these beings came from Mars—not the Mars that we now know or that existed then, but the Mars of approximately one million years ago. At that time it was a beautiful planet, totally alive, not dead in the way it has become. But the inhabitants were suffering from, and in fact the planet was being destroyed by, the effects of a "Lucifer rebellion" stimulated by the same type of sickness we would later encounter on Earth. But the Martian problem was not created by Lucifer himself— rather from a similar type of character. I'll call this general problem "the Lucifer rebellion," even though Lucifer himself was involved in the most recent upheaval only.

The Lucifer Rebellion

Lucifer and Michael were two of the most incredible angels that God ever made; they understood the reality from beginning to end. The only being that was beyond them was God Himself. Lucifer and Michael chose different paths. Michael remained connected to life, and to God; he remained connected to the formless and connected to the Light.

Lucifer in his attempt to ascend to the heights of God, began to create a separate reality. He created an external Merkaba—a synthetic field that was a space-time interdimensional vehicle—and was able to move through all space, time, and dimensions. This is what we call a craft or a flying saucer. He was the originator of all of this.

Creating it internally as a living field meant that you had to have your emotional body intact along with your mental body, which protected you. Lucifer, by going one step further and creating it externally, separated himself from God, and was no longer able to do it internally.

Now this type of experiment against God had been attempted three times previously, and it always ended in total chaos. The fourth and most recent Lucifer rebellion was about two hundred thousand years ago, and at the time he convinced about a third of the angels to join him.

About a million years ago the race on Mars was dying from the effects of an earlier Lucifer rebellion (the third one). The planet was terminating from Merkabas run amok. When you create a Merkaba internally using love or your emotional body, it becomes a living field around your body, but when you create it externally you don't have to use love—you only have to use the calculating mind. Ultimately, this act produces a being with only a left brain who doesn't have an emotional body or understand love. The best example of such a race is the Greys, descendants of the Martians and one of the alien races now visiting the Earth. Another effect of creating the Merkaba externally is that the act itself is generative of duality. How could it not be, since it dualizes in order to externalize and turns emotions into technology? It then becomes increasingly difficult to perceive the One Spirit that moves through everything. So, we see good and evil and, even though the One Spirit is still present in an externally created world, it is incredibly difficult to discern.

When the Martians came to Atlantis they imported the effects of the Lucifer rebellion right along with them, and this is the deed that led to our downfall on Earth. The problem was that Mars was a full left-brain culture; the Martians knew and understood everything intellectually but they had no feeling—specifically, they had no love. They had no reason to care for anyone other than themselves. As a consequence, they were always fighting, until finally, they destroyed their atmosphere. As Mars was dying a small group of Martians, approximately a thousand or so, built structures in the region that we have come to call Cydonia—the massive humanoid face and the "monuments" that the

Viking spacecraft photographed on Mars in 1976. These monuments represent in exquisite mathematical detail at many levels the form of a star tetrahedron inscribed in a sphere, and they also describe how the Martians created their external Merkaba. In fact, these monuments—a vast series of pyramidal structures—became the actual vehicle in which the Martians transported themselves. An external Merkaba can be created in either of two ways—either in the more familiar shape of a UFO, or in a building or series of structures. It was an external field because they had lost the ability to create an internal Merkaba; after all, such a vehicle requires an emotional body. They certainly knew how to create an external field, and they did just that. They got in it and left, believing this to be their only option.

Through the creation of this elegant external time-space vehicle, the Martians were able to journey into space, travel in time, and discover and decide the perfect time and place for them to enter. They saw, projected far ahead into their future—about sixty-five thousand years in our past—this place on planet Earth on Atlantis, so perfect and waiting for them. That is where they headed. They arrived without permission and tried to take over immediately, but there were too few of them and they failed. They finally decided to go through the motions of trying our feminine way even though they didn't understand or accept it; but they put on a good act. They tried it, in fact, for fifty thousand years but always with disruptions. Just imagine being stuck in a bad marriage for fifty thousand years! Their influence on us was so strong that we began to switch our consciousness from a female to a male orientation. We didn't transmute completely, but we were significantly distorted.

When the Martians stepped into our evolutionary pattern in Atlantis, we were about the equivalent of a thirteen or fourteen-year-old female and they were the equivalent of a sixty-five-year-old man. They came in against our will and, essentially, raped us. As I said, they would have taken over if they could have, but there weren't enough of them so they had to go along with our ingenue program, at least for a while.

Even though there were always conflicts, things progressed fairly smoothly until about sixteen thousand years ago. At this time a comet hit planet Earth where Charleston, South Carolina, is today. Remnants of the molten stone scattered over an area the size of four states, making huge impacts. The Atlantians were an extremely advanced culture at that time, so they knew the comet was coming. In fact, they had the technology to blow it out of the sky. The Martians advocated using the machinery to destroy the intruder, but even though they had a strong influence on us, our female orientation was still strong enough for us to say no. Basically, the intuitive aspect said, "No, don't shoot it out of the sky: it's God and what will be will be, so let's just let it happen." So, the Atlantians watched the comet hit and, guess what, most of the damage occurred right in the area where the Martians were staying. In addition to many deaths, most of that area totally sank. For the remaining Martians, that was the last straw.

They decided from that time on they would not follow our lead anymore and would go their own way. What they did was to assemble another Luciferian experiment just like the one carried out a million years ago on Mars that resulted in the external vehicle. They still did not have the requisite emotional body and love necessary to create counter-rotating fields as a living entity, for they had done nothing to nurture this aspect of themselves on Earth. They had, or at least they thought they had, the ability to reinvent the Merkaba externally. In so attempting to create it, they intensified the pathology of the Lucifer rebellion on Earth and they also failed miserably. The experiment lurched out of control and began to rip open the dimensional levels, causing spirits who were not meant to be here—spirits who had been consigned to other dimensional worlds—to come in by the millions. This blew the lid off things; the place just got crazy.

The ascended masters helped a great deal by sealing up most of the dimensional tear as quickly as they could, but with millions of these disembodied beings already present, every Atlantian had anywhere from twenty to a hundred spirits in his or her body. Things in Atlantis got much worse than they are here now, though we are fast approaching that.

This failed experiment happened about sixteen thousand years ago, and for the next four thousand years matters just kept worsening. The ascended masters, our highest aspect, the very highest degree to which consciousness on this planet has reached, prayed for help. They had to find a way to rescue everyone. They couldn't just get rid of the Martians or kill the disembodied spirits; that is not what ascended masters do, it is not the path of life. So what they were looking and praying for was help in finding a win-win situation for everyone.

In the highest aspects of life there is only oneness, and in Unity consciousness there is no duality, no illusion. It is abundantly clear that One Spirit moves throughout the entire creation, and that everything is a smaller part of the whole. This is precisely why the ascended masters couldn't just obliterate the Martians. Our modern surgical philosophy of excision—to cut you up and take out what isn't working—is not what true life is about. By annihilating the Martians or driving them off, a loss for anyone would ultimately have been a loss for everyone.

A lot of intergalactic consuls were involved in the ensuing deliberations. They looked back into the remote past and were able to see something that had been done before and in fact had worked. It required initiating planetary Christ-consciousness—not merely allowing it to proceed naturally, but synthetically beginning the process of creating the Christ grid around the planet, with the result that everything would eventually be healed. Once our awareness reaches a certain level, all the problems solve themselves. Christ-consciousness is Unity consciousness, so if it can be achieved, everyone will win.

To be clear, permission to proceed was granted only because the ascended masters, our highest aspect, had already reached this level. However, the grid for Christ-consciousness did not survive the ripping open of the dimensional levels.

There are five levels of consciousness associated with planet Earth. These levels are directly related to the number of chromosomes we have in our genetic makeup. They also each prescribe a range of height. The first level has forty-two plus two chromosomes, and this is harmonic with Unity consciousness. At this point collective consciousness operates such that if one person experiences something, it is possible for everyone else to access this memory and relive it in total detail through the holographic re-creation of the event. This is the dreamtime of the Aborigines in Australia. The range of height associated with this level is three and one-half feet to five feet.

The second level is where we are now. We no longer have this Unity consciousness; we are cut off and separated. We have forty-four plus two chromosomes, and our approximate range of height is from five feet to seven feet.

In the third level, which is Christ-consciousness, there are forty-six plus two chromosomes. The height range is from ten to sixteen feet. Here we are back into unity memory again, but its form in the third level is upgraded into one of instant manifestation; it is no longer dreamtime but real-time. When you remember something, it is real. It is not just your memory but the memory of all Christ-conscious beings who have ever lived. At the third level there is really only one consciousness that moves through everything; its key is memory. This is what immortality is all about. Immortality is not living in a body forever because there is always some place higher to move. The key is not having a break in consciousness as you move through the different levels, not having a memory loss, being able to leave when you want and continuing to know where you've been.

The fourth level of consciousness has forty-eight plus two chromosomes and a height range of twenty-five to thirty-five feet. The fifth level has fifty plus two chromosomes and a height range of fifty to sixty feet. The fourth level is disharmonic like the second but it is a necessary step to get to the fifth and highest level that can be achieved on this planet.

Thoth

Thoth is a specific historical man who went through ascension fifty-two thousand years ago. For sixteen thousand years he was the king of Atlantis, where his name was Chiquetet Arlich Vomalites. He remained on Earth in the same body until May 4, 1991. He could have left earlier—most ascended masters have—but he was among a small group who decided to stay. Knowing without any doubt whatsoever that all things are interconnected and that there is One Spirit that moves through everything, Thoth preferred to remain here as a teacher.

He did leave for about two thousand years and traveled to various planets. He would arrive at a planet and sit there for a hundred years or so, observing and learning how the inhabitants did things. Then he would go to another planet, until he finally returned here. His vow was to remain on Earth until we reached a certain level of consciousness. We have now reached that level, so he left this planet on May 4, 1991.

Evidently, what happened before, during, and after the Gulf War was a culmination of something. The light on the planet is now stronger than the darkness for the first time in sixteen thousand years. Even though we don't see it yet, the power balance has shifted and the laws have reversed themselves. Now when negativity resists the light, which is its very nature, instead of overpowering the light it gives more power to the light and we get stronger. So, hang on!

Thoth's most influential act was introducing writing to the planet. He was called "the scribe" in Egypt, for he is the one who wrote down all the ancient history. That is why Drunvalo was sent to him. Most of Drunvalo's information about us and our history comes from Thoth, who was always quick to point out that he may not have it one hundred percent accurate, but his account is probably pretty close to what actually happened.

Drunvalo first met Thoth in 1972. He was studying alchemy—that is, how to turn mercury or lead into gold—not for the purpose of making money but to observe chemical reactions. All chemical reactions have parallels in life somewhere, on one level or another. By understanding chemistry and the way atoms combine to form molecules, and how these molecules recombine, you can see in tiny detail how larger operations happen. True alchemy is primarily understanding how our level of consciousness goes into Christ-consciousness.

Drunvalo was studying this system with a master. One day they were doing an open-eye meditation. After about an hour, the teacher disappeared from the room. In two to three minutes a completely different body appeared in his place. This person was short, about five feet, three inches, and he looked about seventy years old. His appearance was ancient Egyptian, and he wore very simple clothing. Drunvalo especially remembers his eyes, which were just like a baby's, very soft with no judgment.

This being told Drunvalo that there were three atoms missing in the universe, and he wanted him to find them. Drunvalo had an experience, which he will not describe, in which he understood what was meant. The being bowed, said thank you, and disappeared. A few minutes later the alchemy teacher reappeared. He knew nothing of this; in fact, he thought he had been there the whole time.

Drunvalo didn't know then that the person who appeared to him was Thoth, and he didn't see him again until November 1, 1984. At that time, they began to communicate regularly over the next few years.

Anyway, to get back to the story, Thoth together with Ra and Araragat, who were also former kings of Atlantis, got the blueprint for the Christ-consciousness grid and went to Egypt (which at the time was a tropical rain forest, known as Khem). They went there because the axis for the now inoperative Christ grid exited the Earth there. The plan that came down from higher life forms was to build the new grid on the old axis. Then with the power of intention alone, they created a hole—about one mile deep, and lined it with bricks. This hole went straight down the axis. This axis comes out in Egypt and also on the other side of the Earth in Moorea, a small island near Tahiti. Thoth

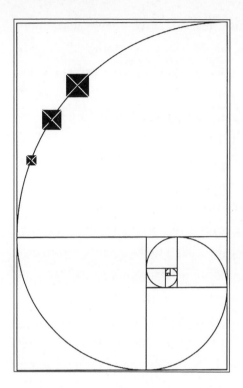

Figure 6.1. The logarithmic spiral on the Giza plateau.

says that there is a vortex or spiral on both ends of the axis, and that a shadow of it formed on the ground would look like a logarithmic spiral (fig. 6.1). They then built three pyramids on that spiral.

The primary purpose of the pyramids was to move our consciousness from the second level (where we are now) to the third or Christ-consciousness level. This was an instrument for planetary initiation, designed specifically to take a forty-four-plus-two-chromosome person into Christ-consciousness and stabilize him or her.

According to Thoth, the pyramids were built with the mind and heart, manifested from memory on the fourth-dimensional level. They were constructed in a period of three days, from the top down. They linked the very consciousness of our evolutionary pattern with the logarithmic spiral. Deep below the pyramids the ascended masters laid out a small temple city that held about ten thousand people and which, by the way, is still there.

Every single life form on the planet has a grid—an electromagnetic, geometrically shaped energy field—connected with it. Even if that species exists on only one spot on the Earth, its grid extends around the whole planet. These grids subtend an average of sixty feet from inside the Earth to about sixty miles above it. If you were to see them superimposed over each other it would look like a light, white-blue haze coming off the Earth.

The most intelligent, advanced, oldest life form on this planet is the whales. Next are the dolphins, then humans. We think that we are the most advanced, but the whales and dolphins are far beyond us. We think we are the most advanced because we can create external things. That, however, is an aspect of the Lucifer rebellion brought from Mars. The most advanced life forms do not create externally; they create everything they need internally.

The whales have been on this planet, conscious and alive, for five hundred million years. They hold the memory of the planet. This is what the movie *Star Trek IV* was all about. Without the whales we would have no memory, we would be lost.

The dolphins have been around here for at least thirty-five million years. They even came out and walked on land for a while and decided against it. These creatures are amazingly aligned with humans. They are mammals, not fish, and both sides of their brains function at one hundred percent. When they sleep, they turn off half of their brain. We have only half of our brain working at any time; the other half is shut down. Of the half that is on, only about five to ten percent is used. So, from the dolphins' perspective, we are not only asleep, we are also unconscious.

There are three grids for human consciousness around the planet. The first one is forty-two plus two; the one we are in now is forty-four plus two; and, as of February 4, 1989, there is also the third grid, which is the Christ-consciousness grid of forty-six plus two. Without the grid, no planet can go into that level.

This then is what Thoth, Ra, and Araragat were preparing. They were beginning to consciously manipulate the internal energy flow of the Earth in such a way that it ultimately would dramatically alter the flow above the Earth. This is a process known as geomancy. This they hoped, would ultimately result in the synthetic creation of the Christ-consciousness grid, which would, in turn, give us the vehicle to move into the next higher level of awareness.

So, this trio made a hole aligned directly with the axis of the old Christ grid. Then they laid out three pyramids, a major project of geomancy. Subsequently they situated eighty-three thousand sacred sites throughout the planet. These were created totally on the fourth-dimensional level. Then over a period of thirteen thousand years they drew humans from every race and all walks of life to build the requisite church here and pyramid there to establish an operational pattern on the third dimension. Scientists may yet discover that all the sacred sites on the planet are laid out in either logarithmic or Fibonacci spirals, mathematically connected and delineated back to that single spot in Egypt. In fact, there's one man who apparently has done just that—his name is Carl Munck. Consider the following from an article entitled *"The Code" of Carl Munck and Ancient Gematrian Numbers.*

> The great mysteries of life are quite elusive. We do not have the "hard facts" needed to feel sure that our theories about the mysteries are true. Sometimes we feel sure, but convincing others is not so easy. Alas, they want "facts," and we cannot produce them. Well, times are changing.
>
> This is the start of a series of articles that will present many "facts" concerning some major mysteries of our world. These "facts" will show evidence that:
>
> - The ancient sites around the world are very precisely positioned on a global coordinate system in relation to the position of the Great Pyramid at Giza.

- The positions of the sites are given in the geometry of their construction.
- A very ancient system of numbers was used in the system, which we will call "Gematria."
- The Nazca Line ground markings "locate themselves" on The Code Matrix system.
- Crop circle formations suggest the same ancient numbers by way of their positions and measurements.
- The very ancient "Monuments on Mars," including "The Face on Mars," were positioned in exact locations, just as the ancient sites on Earth.[7]

And there is this from *The Mayan Ouroboros*. The following is quoted directly from the book:

Going further, what if you study upwards of 250 Sacred Sites and find that the longitude relative to Giza and the exact latitude are encoded into every single one? Even if there were only 10 sites involved, you would have ruled out coincidence. And Carl Munck has discovered that you can find the exact longitude relative to Giza and the latitude, down to several decimal places, encoded into every single one of the 250 Sacred Sites that he has studied!

In proving that the placement of the Sacred Sites he's studied could not have occurred by chance, Munck also proved that the people who built them must have been able to view our planet from outer space! It would have been literally impossible for us, today, to verify the accuracy of these ancient builders' calculations before we ourselves had satellites!

So Munck is giving us proof that there was a much higher technology in ancient times than classical historians have assumed. Besides that, the overwhelming implication is that the Sacred Sites—at least the ones that Munck has visited—were planned and executed by the same mind or agency. They were all built according to a single plan.[8]

This area in Egypt, only recently discovered, is now called "the solar cross." The Association for Research and Enlightenment in Virginia considers this to be one of the most critically important places in Egypt.

In addition to what the masters did in Egypt, they bequeathed us the second level of consciousness, the one we are now on, as an intermediate disharmonic step toward the third—or Christ-consciousness—level. They did this through Thoth, by introducing writing, which caused us to lose our dreamtime or unity memory. Prior to this, writing was not necessary because recall was total and instant. The second level of consciousness, the disharmonic stage that we are now in, intervenes because life has not yet figured out how to go directly from the first to the third level. However, the second level is a tier that one wants to get on and off as quickly as possible because if a civilization stays on it too long, its planet runs a high risk of destruction. We can't get off this level any too soon.

The complex in Egypt I've been describing was built about two hundred years before the deluge and the shifting of the poles. This, by the way, was *the* flood, the one in which Noah floated off in the ark and survived. Immediately prior to the shifting of the axis and the deluge—after two hundred years of patiently waiting for the telltale signs—Thoth flew to the Sphinx. This marks the oldest object on the planet which, in actuality, lies about one mile beneath the surface and is a very large spaceship. Thoth presented this along with much of the information found in this chapter in an ancient document called *The Emerald Tablets*.

According to Thoth, this ship is used to protect us. He says every time we approach a pole shift—the duration of which is about twenty hours—we become extremely vulnerable because as the magnetic field of the planet collapses, we go through a three-and-a-half-day period of total darkness as we enter the void between dimensions. At this time the dark side always comes in and tries to dominate. This has happened like clockwork throughout our 5.5-million-year history, but each time, one very pure person has found the ship and lifted it into the air, and whatever this person thinks or feels happens. It must be a person who has stepped into Christ-consciousness, so that what they think and feel in fact does manifest instantly. This act always prevents the dark side from taking over.

As we are again approaching a pole shift, the spaceship has already been put into place. In 1989, a woman from Peru crossed into Christ-consciousness, raised the ship, and thought the following: the Greys are suffering from a terminal illness found only on Earth. And remarkably, this is what immediately began to happen. By the end of 1992 the Greys, to a one, were gone. Their only recourse was to get out of here; they couldn't live on Earth anymore.

Back to the craft—I don't want you to have any misunderstanding. In his 1925 translation of *The Emerald Tablets of Thoth, The Atlantian*, Doreal said the ship was powered by atomic motors. Well, that wasn't even close; in fact, it is designed to attach to and run off of your own Merkaba. Furthermore, this spaceship is only three to five atoms thick, and it is flat on the top and bottom, about two city blocks across, and round. Usually, it is one overtone higher than whatever the Earth is on, so even though it is one mile beneath the planet's surface, it can easily pass right through the Earth.

After the Martians' failed experiment on Atlantis, there were approximately four thousand years of life on Earth getting more and more chaotic. Then as we reached the point in the precession of the equinoxes where the shifting of the poles was imminent, Thoth swung into action. He raised the ship, flew back to the island of Undal, and picked up about sixteen hundred ascended masters. They had gotten no more than a quarter of a mile off the ground, he said, when Undal sank. That was the last part of Atlantis to go under. The ship then traveled back to the Great Pyramid as the magnetic field of the Earth collapsed. If this collapse lasts for fourteen days or longer, it takes our memory with it. Our collective memory is directly dependent upon the magnetic field of the Earth, so if it collapses, we have no idea who we are. But if you have mastered the Merkaba you can create your own magnetic field from the counter-rotating fields of light and retain memory.

There were survivors of Atlantis then, but all of them with the exception of a few—the Mayans and the Kogi, amongst others—had completely lost their memory. Just imagine for a moment what this must have been like. If all of a sudden, your memory was gone, what good would anything in our high-tech modern world do you? How about your car, when you wouldn't even remember what your keys were? It's back to square one.

The masters landed on top of the Great Pyramid, which was constructed in such a way that it created a perfect landing platform for this spaceship. There they formed a living group Merkaba field from which a large counter-rotating field of energy extended 1.6 million miles into space. For the critical 3.5 day period during the pole shift they controlled the axis, the tilt, and the orbit of the planet. In fact, they changed the orbit; it used to be a 360-day periodicity, and now it is 365 and a quarter days.

They stayed in the ship for the period in which the magnetic field was collapsed, and then they found a whole new world. Atlantis was gone, portions of what is now the United States had risen above water, and the planet was on a different, much denser vibratory level.

Ra along with about one-third of the masters entered the Great Pyramid by a circular tunnel leading to its underground city. Tat, Thoth's son, was among them. This group later formed the Tat Brotherhood, and there is a large community of immortal beings living there to this very day.

The ship then flew to Lake Titicaca and the Island of the Sun. There Thoth disembarked with another third of the masters, and they founded the Inca Empire. The ship next journeyed to the Himalaya where Araragat got off along with a little less than one-third of the masters. The rest of the ascended masters (seven of them) returned to the Sphinx, raised the ship in its overtone so it could pass through the Earth, and then descended about a mile below the surface into a circular room where the ship itself remained until 1989.

These three places—the underground city, the Island of the Sun, and the Himalaya—were chosen for very specific reasons having to do with planetary geomancy as set up by the masters for a synthetic Christ-consciousness grid of the Earth. The Egyptian aspect became the male point of the grid; the Mayan-Incan aspect became the female counterpoint of the grid; and the Himalayan aspect became the neutral or child point of the grid.

Egypt and Stair-Step Evolution

Egypt became the home not only of the Tat Brotherhood, but also many of the survivors of Atlantis. With their memory erased by the pole shift, the Atlantians reverted to barbarism, reduced to the basic survival skills of building a fire to stay warm and so on. They had to wait a long, long time until they could even begin to develop again. In fact, it wasn't until about 3800 BCE that development picked up, when the Nefilim began to reestablish their terrestrial connections in the place where they had set up their original bases in southern Iraq. The Nefilim simply gave back the lost information. As a result, Sumer was born.

There is a clear discrepancy between what Thoth says and the writings of Sitchin regarding the development of the Egyptian civilization. Sitchin believes that the Sumerians brought their culture to Egypt, but Thoth says no—it was our own ascended masters, the Tat Brotherhood, who established the Egyptian civilization.

In both Sumer and Egypt, then, there was an amazing correspondence. Each culture came out in its fullest and finest form virtually overnight. Then both cultures began to degenerate from there. Attaining full bloom overnight is roughly equivalent to a modern-day airliner suddenly appearing in 1903 with no prototype. Archaeologists have absolutely no explanation. Sitchin, in *The 12th Planet,* calls Sumer "The Sudden Civilization."

The Tat Brotherhood closely monitored the Egyptians. When they felt the time was right, they sent out a small group—either one, two, or three of their own members, dressed exactly like the Egyptians—and began to re-seed the knowledge of Atlantis. This is called stair-step evolution. There are no evolutionary patterns; all of a sudden the people just know everything about a certain subject. Then there is a little plateau, and all of a sudden, they know everything about another subject, and so on.

As soon as a particular piece of information was given it would almost immediately begin to degenerate. The explanation for this lies in the precession of the equinoxes. As we move away from the center of the galaxy in this twenty-six-thousand-year cycle, we fall asleep. After the last pole shift the Earth was at the point in the precession where planetary consciousness must fall asleep. Thus, each time new information was given, the people almost immediately began to lose it until about 500 BCE, by which time the Egyptian civilization was almost totally gone.

Akhenaten

The Egyptians were also losing the idea of the One Spirit and began to worship many gods. In order to explain this, we have to go back into Atlantis for a moment. The Naacal Mystery School was reestablished on Unal, where Ay and Tiya, along with the thousand graduates continued to teach immortality. The only problem is, due to memory loss from Lemuria or whatever, it took more than twenty thousand years for the first person to finally get it. His name was Osiris—even though the story of Osiris is told as though it happened in Egypt, Thoth said it was an actual lived event in Atlantis. In addition to being overjoyed, the Naacals used Osiris as the actual example for others to follow. And since the Atlanteans had holographic memory, all of this was transmitted orally.

In Egypt, where we no longer had total recall, it was written down in *The Forty-Two Books of Thoth*, with two additional books set aside. What it was—since the understanding of ascending into immortality was now in the chromosomes of Osiris—was the "mapping of movements" of the 42+2 chromosomes in the void that enabled one to move from the first level into the second level of consciousness. This was the necessary preparation for moving into the third level—the immortal level of Christ-consciousness. Each chromosome was represented by strange looking animal-headed creatures called *Neters*.

While in Atlantis, since we were in Unity consciousness, it was clearly understood just what these Neters represented. But in Egypt because we were now in the separate state, of the second level of consciousness and no longer saw unity, these Neters became worshipped as gods. And since there were forty-two plus two of them, there were now forty-four "gods" being worshipped in the Egyptian religions. But wait a minute, when you realize that Upper Egypt had a slightly different interpretation from Lower Egypt, it was now eighty-eight different "gods" that were being worshipped.

The ascended masters saw this as a huge problem, so serious that if nothing was done to correct it, we would not have survived. So, they addressed this problem with another direct intervention. They decided to have an actual Christ-consciousness being walk on the very surface of the Earth to put the real thing back into the Akashic records.

This was the person we know historically as Akhenaten (fig. 6.2), who, by the way, was fourteen and one-half feet tall. He was not from the Earth but from the star system of Sirius. He developed a whole new religion, the religion of the sun. That is, the sun was worshipped as a Unity image.

Akhenaten, after the construction of his city of Tel el Amarna, was given only seventeen and a half years around 1355 BCE to make his

Figure 6.2. Akhenaten.
"The Coronation of Akhenaten" from
Akhenaton: The Extraterrestial King *by Daniel Blair Stewart.*

imprint. In the meantime—with the exception a select few, who were to become his students—almost everyone hated him. He disrupted all the religions, telling people that the priests were not necessary, that God was within each person, and that all they needed to do was learn how to breathe and everything would be fine. He told them henceforth there would only be one religion for the whole of Egypt, and no one wanted to hear that. Even though Egypt had the strongest army in the world, Akhenaton, who was a pacifist, told them they couldn't fight anymore. He ordered them to stay within their borders and respond only if attacked.

Akhenaton gave initiates a twelve-year advanced training in the "missing knowledge" (I will describe this school in more detail later). This course produced almost three hundred Christ-consciousness beings, most of whom were women. From 1350 BCE until roughly

500 BCE, they joined with the Tat Brotherhood and remained in the underground city beneath the Great Pyramid. Then they migrated to Masada, Israel, where they became known as the Essene Brotherhood. Mary was one of these immortal beings and was thus in the "inner" circle. Joseph, by the way was from the "outer" circle of the Essene Brotherhood and was not immortal, but he was absolutely necessary. So the next step began, which was to bring forth Jesus to demonstrate the actual process of moving from human-ness into immortality. To be clear, Akhenaten was the living example of what we would become; someone however had to do it—that is someone had to be the example of how to move from the second level of consciousness into the immortal state of the third level—and get it into the Akashic records. That someone was Jesus.

The Egyptians disposed of Akhenaton after a seventeen-and-a-half-year reign. Then they did what they could to erase the memory of him, including completely raising his city of Tel el Amarna. Everything reverted back to the old ways. In spite of this, he was ultimately successful. He wasn't after a lasting legacy; all he needed to do was to get his example into the Akashic records, the living memory of the Earth. He needed to establish the Essene Brotherhood, which would suffice to get the next stage going. He did exactly what he was supposed to do in the time he was allotted.

7
Melchizedek

As I mentioned in chapter 5, if you knew how to change your consciousness wavelength and rotate by ninety degrees, you would discover your innate ability to travel inter-dimensionally. You would rapidly go through the colors of the rainbow, and then disappear from this reality in a ball of light. You would end up in whatever dimensional world you had tuned to, and you would be just as real there as you were here.

However, each dimensional level is completely different, and you must be consciously mature enough to handle the higher—or lower vibrations—of the level you have just entered into. If not, you will be sent right back to the world from which you came.

There are certain beings—perhaps many of them—who have progressed to the point to where they are able to go anywhere and remain conscious and stable.

The act of learning how to move through all 144 dimensional levels (the 12 dimensions and the harmonic overtones of each of them) teaches beings how to go across the great wall and into the next octave. The person then comes out the other side and sees the big picture, with the realization that all of creation repeats itself over and over.

At that point there is a decision to make—to go beyond all of creation and back to the source that is outside of this created reality, or to remain here. If the person remains, he or she is called Melchizedek. Life then uses them to heal interdimensional problems. They all have one interpretation of the reality; they understand there's One Spirit moving through them. They are really like cells in a larger body.

We have all been around since the very beginning, and we will all be around until the end, so we have all been there at one time or another. We have all been at the highest levels, and at the lowest—with many stops in between!

Let's now consider the possibility that there are perhaps ten million or so Melchizedeks in the galaxy, one of whom is on planet Earth. According to Drunvalo, Machiavinda Melchizedek is the person who was assigned from Galactic Center to be with us from the moment of our inception.

The Great White Brotherhood and what you might call the Great Dark Brotherhood are two bodies of consciousness opposed to each other in every imaginable way. Machiavinda—as are all Melchizedeks—is from the Great White Brotherhood, a body of beings, of which there are seventy-two orders, that does everything it can to advance our evolution, while its dark counterpart does everything it can to induce fear and delay evolution.

These seemingly opposing forces tend to balance each other out so that our evolution takes place at exactly the right time, neither too soon nor too late. Viewed from a higher level—the fourth dimension or above—this is Unity consciousness. The two are just different aspects of the One Spirit working in harmony. It is only because we are down here in the midst of polarity consciousness that we see it in terms of good and evil.

Due to amazing events that happened in 1972 (I will devote a whole chapter to this later), the Great Dark Brotherhood, who, by the way, knew what was going to happen, enlisted four additional members from the star systems of Orion, while the Great White Brotherhood sent four of their own in response to this.

Now, let's see if we can track the descent of a Melchizedek (perhaps you) from the heights of the thirteenth-dimensional realm, down to the level of the third dimension, where self-aware consciousness first appears. You would begin by placing a veil over your memory of the thirteenth dimension. Why? Because, to have memory now of life there would be just too painful. You could not exist here in the denseness of the third and maintain full memory of what that level of life is like.

Your descent down through the many levels would of necessity, be a slow one, to give you time to acclimate. From the thirteenth dimension, you would be given a movement pattern, and use it to traverse through the Great Void. You would maintain this pattern for a long time, perhaps for millions of years. However, since time in the higher worlds is spherical, it may also be for only a moment, Regardless, you would continue to move until light reappeared. If your final destination was third-dimensional Earth, you would be met by Machiavinda. Then you would go through the center of a nebula, the middle star in the belt of Orion. This is one of the primary star-gates to other dimensional levels. For instance, there are thirteen different star-gates in our galaxy, but the middle star in the belt of Orion is a special one. At this star-gate, great light and great darkness operate very close together. Many of the Greys emanate from precisely this part of the galaxy.

After going through the belt of Orion, because the Pleiades is a galactic university, you would most likely go there, to a planet where its inhabitants' dwell on the higher overtones of the fourth dimension, and all learning is accomplished through pleasure and joy. All teaching utilizes games.

As you continue to wind your way slowly down through the dimensional levels, you would likely make your way to the third planet out from Sirius B, to a world that is almost all ocean. The Sirians are also on the fourth dimension, but on a lower overtone. They do not yet experience joy and pleasure to the same degree as the Pleiadians, but they are getting there. On this marine planet, who better to connect with than an Orca whale. Since the whales hold the memory pattern of the 500-million-year history of the Earth, you would be thoroughly brought up to speed.

When it's time to move on, let's suppose you were taken to the land mass and given an already-made adult Sirian body, whose cells contain the memory patterns of how to run the Sirian ship that was also given to you, along with a full crew, and a prepared flight pattern for Earth. You would fly from Sirius B right through the middle of Sirius A. You pass through successfully by tuning to the same vibration as the sun so that "hot" is no longer hot. Ninety seconds later you come out through our sun. This is because of our intimate connection with Sirius. Using this maneuver, you then reached the orbital field of Venus, the world containing the Hathor race, the most advanced beings in this solar system. Venus in fact is the "headquarters" for our solar system. You must first get permission from the Hathors in order to enter.

Then, as you entered Earth's atmosphere, you would leave your Sirian body and ignite into a ball of light, so you could study and learn from the ascended masters, whom you found dwelling on the higher overtones of the sixth dimension. You would find them ready and willing to assist you, and prepare you as best they could, for the harsh reality of polarity consciousness and third-dimensional Earth.

You would then either be born into a baby's body, or walk into the body of an adult. If you are a "walk-in," you would do so by permission from the highest levels; you would enter into your new body in one breath, as the other spirit would also leave in that same breath. Since this was all agreed upon from the levels of Galactic Command, the person who left would be given in return, something very special— something that would ensure an appropriate level of spiritual growth. That person also undertook certain training and schooling that would soon come in rather handy.

Sounds incredible, right? And of course, this story is so fantastic that it couldn't possibly be true—but what if it is? Let's consider the possibility that most of us are not from here; our point of origin is somewhere other than planet Earth. We are extraterrestrial masters, multidimensional beings, having a human experience. There are tens of millions of ETs on this plane. Some have been here for a very long time; others are more recent arrivals. They are all lending their consciousness to the planets' evolution; some are more aware of their role than others.

The value in knowing whether you are an extraterrestrial or not, is in expanding your picture of reality. If you are a human, then you are a being who has lost your spirituality and must somehow regain it. If you are an ET then you are a spiritual force in the universe that is channeling itself through this human embodiment to serve the planet. The entire universe rearranges itself to accommodate your picture of reality. Your identity, your sense of who you are is the major element in your picture of reality, so which one would you prefer?

Remember then, that on the fourth dimension you are a Christ-consciousness being, and you came here to bring your love, light, and wisdom to assist the planet in its transition.

8

Introduction to Sacred Geometry

When the teachings of geometry are used to show the ancient truth that all life emerges from the same blueprint, it is called sacred geometry. These teachings clearly show us that all life springs from the same source—the One Spirit moving throughout all of life. When geometry is used to explore this great truth, a broader understanding of the universe unfolds and we can begin to see that all aspects of reality are sacred. Understanding the simple truths of sacred geometry leads to an evolution of consciousness and an opening of the heart that is the next step in the process of human evolution.

Once available only to secret mystery schools, the teachings of sacred geometry are now available to everyone. They can be used to help you connect more fully with the universe, as well as assisting with emotional and physical healing, and greater peace of mind. Once these truths are understood by the mind and experienced through the heart, a whole new world emerges.

Sacred geometry is the morphogenetic structure behind reality itself, underlying even mathematics. Most physicists and mathematicians think that numbers are the prime language of reality, but it is actually shape that generates all the laws of physics.

Figure 8.1. The flower of life.

Sacred geometry is the emblem of reality throughout the cosmos. It is sometimes called "the language of light" and sometimes "the language of silence." In fact, it is a language, a language through which everything was created. The image of the flower of life is the penultimate shape in sacred geometry (fig. 8.1). Contained within the flower of life is everything within this wave form universe. There isn't anything in the reality and never will be that isn't manifested through that image—all languages, all laws of physics, all biological life forms—including all of us individually.

The flower of life can be found in various locations around the planet. No one knows for sure how old this image really is. We can get an approximate idea of a minimum age from one of the oldest temple complexes ever discovered in Egypt, a six-thousand-year-old structure called the middle Osirian temple at Abydos, Egypt.

The middle temple is much lower in elevation than the other two temples; it was buried at the time the pharaoh Seti I was constructing the other two temples at Abydos. It is notably different in its construction. Utilizing large granite blocks and amazing precision, this temple is unlike any other Egyptian temple architecture. The flower of life pattern was placed upon the granite siding of this temple. It was not carved into the granite. It seems to have been burned into the granite or somehow drawn on it with incredible precision.

The flower of life has also been found in Masada Israel, Mount Sinai, and many temples in Japan and China. Recently it has been found in India and in Spain as well.

The image is called "the flower of life" because it originates in a tree. Think of a fruit tree: as it grows, it flowers and then fruits. The fruits fall, and in each one are multiple seeds, and each seed has in it the image of the tree. Contained within the geometry of the flower of life is all of creation.

The seed aspect delineates the first circle and the six circles around it (fig. 8.2). The next image is the tree of life (fig. 8.3). Its image is contained in the seed. When you superimpose the two images (fig. 8.4), the

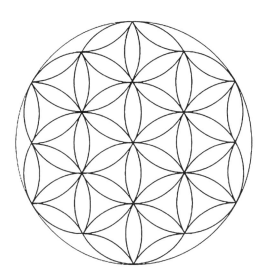

Figure 8.2. The seed of life.

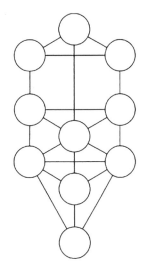

Figure 8.3.
The tree of life.

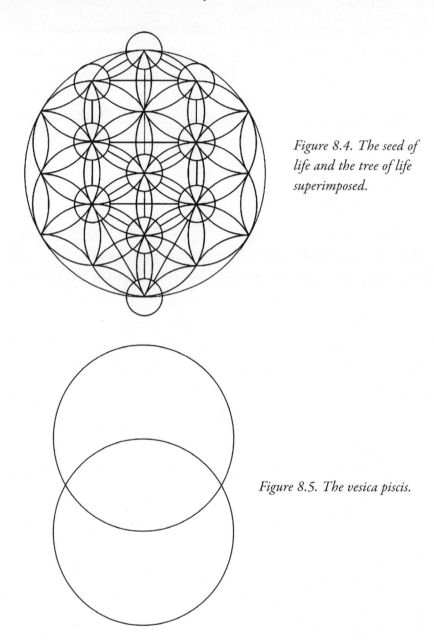

Figure 8.4. The seed of life and the tree of life superimposed.

Figure 8.5. The vesica piscis.

seed of life and the tree of life, you see how every line links and the tree fits perfectly inside.

Another central image in sacred geometry is *the vesica piscis* (fig. 8.5), which is simply a circle next to another circle exactly the same size so that the edge of one circle passes through the center of the other. The common area created is the *vesica piscis.*

The flower of life and the seed of life are nothing but vesica piscis. Also, looking at figure 8.4, you will see that every line in the tree of life is either the length or width of a vesica piscis. The nature of sacred geometry is that it is absolutely flawless—there are no accidents. It continues to unfold until the entire universe is created. Every single part of it—like the reality that mirrors it—is interlinked with everything else. You can start at any point and generate the whole language of creation.

9
The Right Eye of Horus

During his time in Egypt, Akhenaten gathered together a few thousand people, all of whom were at least forty-five years of age. They had all been through twelve previous years of training known as the Left Eye of Horus, which is an emotional body right-brain training. He took these people for another twelve years through his mystery school, *The Law of One*, and gave them the missing knowledge. This was later transmitted to Drunvalo by Thoth.

On a personal note, I made the decision to "enroll" into Akhenaten's mystery school as best I could, So I purchased a high-quality compass and made the drawings—every one of them. And though it was a challenge, it helped to facilitate a huge leap in my understanding of this reality. I learned that the universe is geometric in nature, that all things arise from geometry. Sacred geometry is really kind of like the master key that unlocks all the doors of mystery for us, and gives us a true understanding of the architecture of the universe, how it functions, and what's really going on.

I also went to all the local bookstores in search for anything I could find on Akhenaten. I hit the jackpot when a book entitled *The Law of One, Book 1, The Ra Material*, jumped out at me and screamed "Here I am, read me!" There are five books in the complete set. However, all of it is available on the website: lawofone.org.

The *Law of One* is a series of 106 conversations between Don Elkins, a professor of physics and UFO investigator, and Ra, speaking through Carla Rueckert. These sessions were channeled by L/L Research between 1981 and 1984.

Ra states that it/they are a sixth-density social-memory complex that formed on Venus about 2.6 billion years ago. Ra says that they are "humble messengers of the Law of One, and that they previously tried to spread this message in Egypt through Akhenaten's mystery school, with limited success.

The primary components of the school appear in only one place, underneath the Great Pyramid in a long hallway leading into the Hall of Records. All forty-eight images of the chromosomes of Christ-consciousness are on the upper part of the left-hand side of the wall. This information was otherwise only transmitted orally.

The symbol for Akhenaten's school was the Right Eye of Horus. The right eye is controlled by the left brain, so this is male knowledge; it is the logical side of how everything was created by spirit and nothing—for spirit needs nothing to create the universe.

Following are the first three verses of Genesis, chapter One:

In the beginning God created the heaven and the earth. And the earth was without form, and void; and darkness was upon the face of the deep. And the Spirit of God moved upon the face of the waters. And God said, Let there be light: and there was light.

One thing the Bible left out and which Akhenaton's school very clearly specified was that in order for spirit to move in the Void it had to move relative to something. The Great Void is total nothingness; so, if spirit moved but had no point of reference, how would you know it moved?

The way Akhenaten's school analyzed this was as follows: spirit projected itself out as far as it could go in all six directions—up and

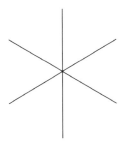

Figure 9.1. Projection of spirit in six directions.

down, forward and backward, and left and right (fig. 9.1). This can be conceived on three axes marked x, y, and z. The amount of projection is irrelevant; even if only one inch, it is enough.

So, spirit projected itself in six directions. Its next step was to connect the lines—first to form a square (fig. 9.2), then to form a pyramid (fig. 9.3), and then to bring the lines down into a pyramid below, which is an octahedron (fig. 9.4). Now spirit had the reality of an octahedron around it. Even though it was just a mental image, movement was now possible because perimeters had been established.

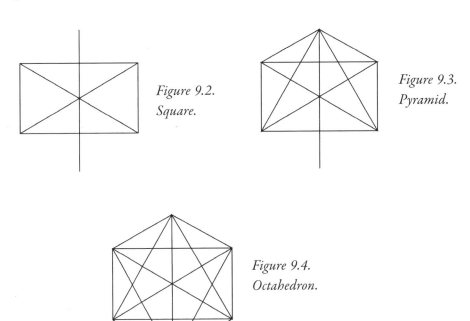

Figure 9.2. Square.

Figure 9.3. Pyramid.

Figure 9.4. Octahedron.

Figure 9.5.
The image of a sphere.

Spirit then began to rotate the three axes, thus tracing the image of a sphere (fig. 9.5). In sacred geometry a straight line is considered male, and any curved line is female. Thus, by rotating the octahedron on its axis, spirit went from male to female, i.e., the sphere. The Bible reports that the male was completed first and the female second. This is a movement from straight lines to curved lines. The reason spirit went from straight lines to curved lines is that the geometric progression necessary for creation is much easier from the female curved lines.

So now the spirit of God finds itself inside a sphere. Genesis says, "The spirit of God moved upon the face of the waters," but where to? In the entire universe there was only one new place and that was the surface. So the students in the school were taught that spirit moved to the surface—anywhere on the surface, it doesn't matter where. The first motion out of the Great Void is to move to the surface (fig. 9.6). After that first motion everything else is automatic; every motion from there on shows you exactly where to make the next motion until the entire universe is created.

The third verse from Genesis is: "And God said, Let there be light: and there was light." After moving to the surface there is only one thing

Figure 9.6.
The first motion out of
the Great Void is to move
to the surface.

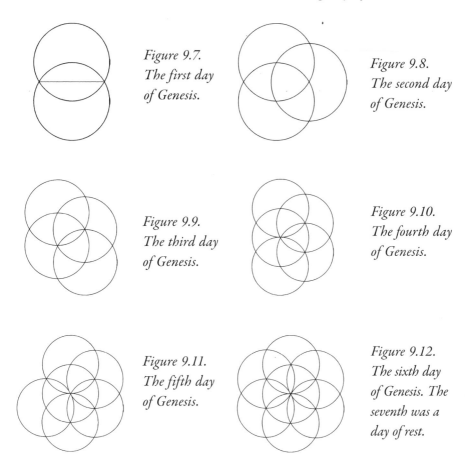

*Figure 9.7.
The first day
of Genesis.*

*Figure 9.8.
The second day
of Genesis.*

*Figure 9.9.
The third day
of Genesis.*

*Figure 9.10.
The fourth day
of Genesis.*

*Figure 9.11.
The fifth day
of Genesis.*

*Figure 9.12.
The sixth day
of Genesis. The
seventh was a
day of rest.*

to do, and that is to make another sphere (fig. 9.7). What you have then is a vesica piscis, or two interlocked spheres, which is the metaphysical structure behind light. And that was the first day of Genesis. Where the two spheres come together is a circle or oval. By moving to this new circle and making another sphere you get the next image, which marks the second day of Genesis (fig. 9.8). Now a rotational motion begins to happen on the surface of the sphere until it completes itself. This is all automatic. See figures 9.9, 9.10, and 9.11.

When you get to the sixth day of Genesis you have six circles fitting perfectly with nothing left over (fig. 9.12). On the seventh day spirit rests, because the genesis and all the laws of the universe are now complete. As

Figure 9.13.
Tube torus.

the image continues to rotate in a vortex, three-dimensional objects start to come out of the pattern.

It is important here to understand that sacred geometry is not just lines on a page; rather, it is the motions of spirit in the Void. It is the map of movements necessary to get out of the three-dimensional Great Void in such a way that, in our case, we end up on planet Earth. Depending on which dimensional overtone you are on, there are actually 144 different forms of voidness.

The first image to come out of this pattern is a tube torus (fig. 9.13). It emerges from the first rotation, or the first six days of Genesis. You create this image by rotating the pattern—when you rotate the pattern, you get a tube torus with an infinitely small hole in the center. Remember, it is a three-dimensional shape, not two. The tube torus (fig. 9.13) is the primal shape of the universe. It is unique in that it moves in on itself; there is no other shape that can do that.

Stan Tenen, through more than twenty years of research, tracked the spiral of a tube torus out of the middle and took out the shape.[1] He removed the minimum amount of matter to delineate the tube torus and placed it inside a three-dimensional tetrahedron (fig. 9.14). He found that by shining a light through it so that the shadow of that shape was projected onto a two-dimensional surface, he could generate

Figure 9.14.
The spiral of a
tube torus inside a
tetrahedron.

©1986 Stan Tenen

all the letters of the Hebrew alphabet, exactly as they are written and in order. He also found that by changing the shape to a different position he could project all the Greek letters. Then by changing the position again he could configure all the Arabic letters. He did this simply by moving this particular shape (the tube torus) to different positions inside a three-dimensional tetrahedron. There are actually twenty-seven primary symmetrical positions inside a tetrahedron.

So, the first thing to come out of Genesis is the connection of metaphysical form to language. And this all occurred during the first seven days of creation.

We have thus begun a rotational vortex energy pattern. Every time a new rotational pattern is completed, a new form is produced, and that new form is the basis of creation. The rotation always begins at the innermost places (fig. 9.15). The next rotation is shown in figure 9.16. By erasing some of the lines in figure 9.16, you will come up with figure 9.17, or "the egg of life." This is a two-dimensional depiction of a three-dimensional figure. The egg of life is actually eight spheres, the eighth lying directly behind the middle sphere. The egg of life is the pattern through which the harmonics of music, as well as that of the electromagnetic spectrum, are connected, and it is also the pattern that underlies all biological life. It is the pattern of all structure, no exceptions whatsoever.

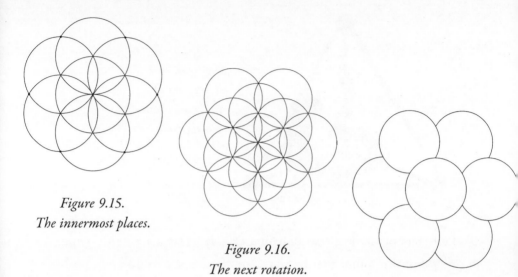

Figure 9.15.
The innermost places.

Figure 9.16.
The next rotation.

Figure 9.17. The egg of life.

The next rotation gives you the outline—that is, the correct number of circles—for the flower of life (fig. 9.18). This flower contains seven circles that just touch each other, as shown in figure 9.19. Figure 9.20 presents how the flower of life is usually depicted. It has been shown this way traditionally because the secret societies that passed it along wanted to hide the next image, which is "the fruit of life." If you look at figure 9.20 you will notice that there are lines and circles that just seem to end, but if you complete all the circles and continue with the rotation (fig, 9.21) you will get to "the fruit of life" (fig. 9.22).

The fruit of life is a very special, very sacred figure. It gives forth the reason for the creation. Thirteen systems of information come out of the fruit of life; we are going to go through one of them here. The full thirteen systems describe in detail every single aspect of our reality, everything that we can think of, see, sense, taste, or smell, right down to the actual atomic structure.

You get to these thirteen systems of information by combining geometric female with male energy. When male and female combine,

Figure 9.18.
The flower of life.

Figure 9.19.
The seven circles in the flower of life.

Figure 9.20.
The flower of life
as it is usually
depicted.

Figure 9.21. To get to "the fruit
of life," after completing all
the incomplete circles, go to the
innermost places and continue
with the rotation.

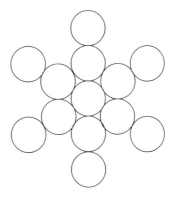

Figure 9.22. The fruit of life.

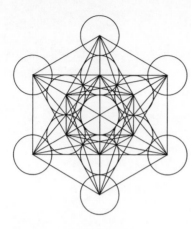

Figure 9.23.
Metatron's Cube.

something new is manifested. Except for the very first form, all the shapes I have been describing have been female-energy curved lines, so one of the simplest and most obvious ways of adding male energy, in straight lines, is to connect all the centers of the spheres on the fruit of life. If you do that, you end up with a figure known as Metatron's Cube (fig. 9.23).

Metatron's Cube contains within it the five Platonic solids (fig. 9.24). These include the cube or the hexahedron, which has six square faces,

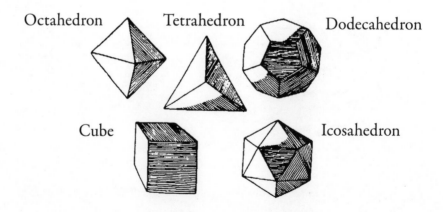

Octahedron   Tetrahedron   Dodecahedron

Cube   Icosahedron

Figure 9.24. The five Platonic solids.

eight corners, and twelve edges; the tetrahedron, which has four triangular faces, four corners, and six edges; the octahedron, which has eight triangular faces, six corners, and twelve edges; the dodecahedron, which has twelve pentagonal faces, twenty corners, and thirty edges; and the icosahedron, which has twenty triangular faces, twelve corners, and thirty edges. The criteria for Platonic solids are that all their edges be equal, that there only be one surface and one angle, and that the points all fit on the surface of a sphere. There are only five shapes known that fit these criteria. The Platonic solids were named after Plato even though Pythagoras used them two hundred years earlier; he called them the perfect solids.

These five figures are of enormous importance. They are the components of the energy fields around our bodies. It is a little-known fact that the five Platonic solids come out of Metatron's Cube. Most of the authors of books on sacred geometry don't seem to understand or realize this fact.

To obtain the Platonic solids from Metatron's Cube you have to erase certain lines. By removing lines in a particular fashion, you first of all come up with the cube shown in figure 9.25. This is a view of the cube on end; it's a two-dimensional image of a three-dimensional object, and it contains a cube within a cube in a very specific ratio.

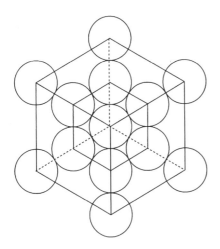

Figure 9.25.
The cube extracted from
Metatron's Cube.

Figure 9.26. The star tetrahedron extracted from Metatron's Cube.

When you erase other lines in a prescribed way, you come up with the tetrahedron shown in figure 9.26. It is actually two tetrahedrons back-to-back, or a star tetrahedron.

Figure 9.27 shows the octahedron, which is back-to-back pyramids; figure 9.28 shows the icosahedron; and figure 9.29 shows the dodecahedron.

In the ancient schools of Egypt and Atlantis these five shapes plus the sphere were also categorized from another point of view. They viewed the elements fire, earth, air, water, and ether as having protean shape. The shapes of the elements corresponded to the Platonic solids as follows: the tetrahedron is fire, the cube is earth, the octahedron is air, the icosahedron is water, and the dodecahedron is ether or prana. The sphere is the void from which all came. So from these six shapes all things can be created.

Atoms, the particles from which matter is formed, are simply spheres with electrons moving around their outer core at nine-tenths the speed of light. This rotation forms an electron cloud, which mimics a sphere. In crystals the different-sized atoms (spheres) align in a straight edge, a triangle, a tetrahedron, a cube, an octahedron, an icosahedron, or a dodecahedron.

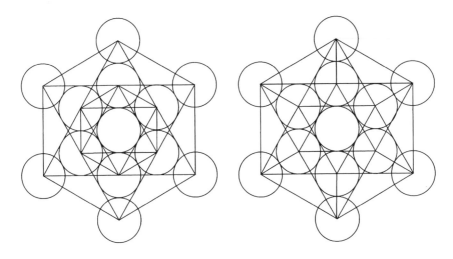

Figure 9.27. The octahedron extracted from Metatron's Cube. (Left)
Figure 9.28. The icosahedron extracted from Metatron's Cube.

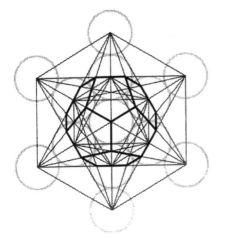

Figure 9.29.
The dodecahedron
extracted from
Metatron's Cube.

Humans

Even though it may not look like it, in fact we are nothing but geometrical images and shapes, both inside and out.

Before conception the ovum is a sphere. It is the largest cell in the human body, two hundred times greater than the average cell. In fact, it is large enough to be seen with the naked eye. So the ovum is one sphere and inside is another sphere, the female pronucleus. It contains half the chromosomes for a human, twenty-two plus one. The membrane (zona pellucida) surrounding the ovum has an inner and outer thickness. In the zona pellucida are two polar bodies.

Conception begins when the sperm reaches the ovum. It takes hundreds of sperm to accomplish this. From these hundreds of sperm eleven, twelve, or thirteen work together as a unit. Through their total unified action, one of the sperm gets to enter the ovum. The sperm tail then breaks off and the head of the sperm forms a sphere exactly the same size as the female pronucleus. These then merge and form a vesica piscis. At that point the two merged cells contain all the knowledge of the universe.

In the next step the sperm and ovum pass right through each other and become cell number one, which is the human zygote. Now it contains forty-four plus two chromosomes. Next, mitosis occurs (fig. 9.30), and the polar bodies migrate to opposite ends of the cell and form a north and south pole. Then a tube forms seemingly out of nowhere. The chromosomes split, with half going to one side of the tube, and half to the other side. The actual proportions of an adult body originate here; there is a "little person" in the original cell.

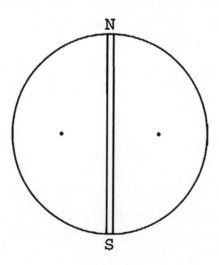

Figure 9.30. Mitosis begins.

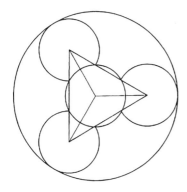

Figure 9.31. The zygote splits into four cells and forms a tetrahedron inside a sphere.

Figure 9.32. The next division yields eight cells and a star tetrahedron.

The zygote then splits into four cells and forms a tetrahedron inside a sphere (fig. 9.31). The next division yields eight cells and a star tetrahedron, which is also a cube (fig. 9.32). At this point, it is the egg of life. The eight cells appear to be identical in all ways and are closer to who we really are than our outer package or body. The location of these eight cells is in the geometrical center of the body, at the perineum, and they are immortal relative to one's body. All the energy fields and grids around a human body are centered on these eight cells. We grow radially out from there.

The first eight cells then divide into eight more cells and form a cube within a cube. That is the last time cellular division is symmetrically geometrical. When you go from sixteen to thirty-two cells, there are two spaces left over, and when you go from thirty-two to sixty-four, it becomes even more asymmetrical. The embryo starts to become hollow and returns to the shape of a sphere. The north pole goes through the hollow ball, grows down, and connects with the south pole, forming a hollow tube in the middle and curling into a tube torus. One end becomes the mouth and the other the anus. From here the characteristics of the particular life form, be it human, animal, insect, or whatever, begin to dominate.

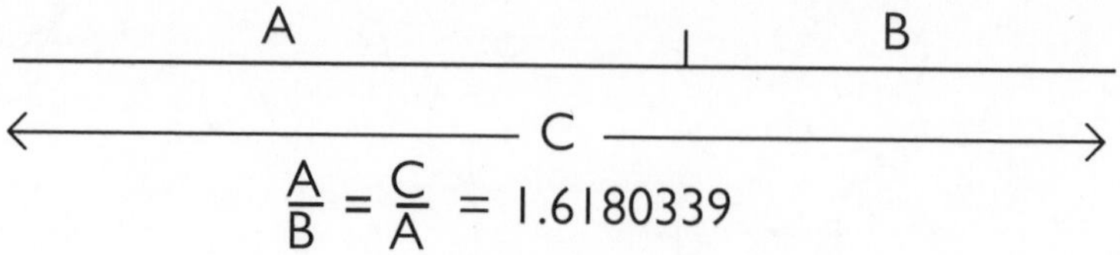

Figure 9.33. Phi ratio.

This is the sequence then: Life begins as an ovum or a sphere, turns into a tetrahedron, then into a star tetrahedron, then into a cube, then into another sphere, and then into a torus.

Phi Ratio

Now let's take a look at the geometry in the space around our bodies. The first concept I want to introduce here is the phi ratio. This is a transcendental number, meaning that it never repeats itself. It is approximated at 1.6180339, but it doesn't end there; it just keeps going on literally forever. The relevance of the phi ratio is that it is found in all known organic structure.

The phi ratio is a proportion. If you have a line (C) and you break C into A and B in a manner that reflects this particular proportion, then A divided by B is equal to C divided by A, or 1.6180.

We can see how the phi ratio is derived by looking at figure 9.34. If you start with a square and then draw a line down its middle as shown in the diagram, then make a diagonal (which is line D in the diagram) and with a compass rotate the diagonal line, then A divided by B is equal to C divided by A, and the proportion comes out to 1.6180339.

The bone structure of organic life is based on the phi ratio. For example, in humans the bones in the fingers all stand in phi ratios. The first bone in the finger is in phi ratio to the second, and the second to the third, etc. This is also true of the bones in the feet and the legs.

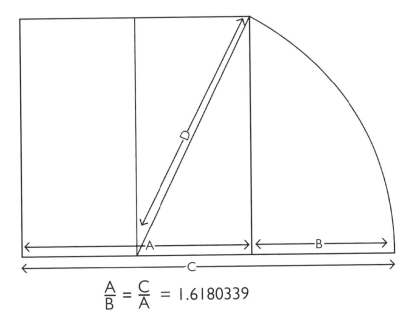

$$\frac{A}{B} = \frac{C}{A} = 1.6180339$$

Figure 9.34. How phi ratio is derived.

All laws are contained in your own body's proportions. The image of the fields around your body is the same image that surrounds everything and through which everything was created.

Look at Leonardo da Vinci's famous drawing, "Proportions of the Human Body" (fig. 9.35). The arms extend straight out and the feet straight down. This forms a square or cube that fits around the body. Its center is located at the base of the spine where the original eight cells are. Those cells also form a tiny cube there. So you have a tiny cube inside the body at the base of the spine, and a bigger cube formed outside (around the body).

When you spread the figure's arms and legs, a sphere or a circle forms with its center at the navel. The circle and the square meet at the feet, and the distance between the navel and the base of the spine is exactly one half the distance from the top of the head to the circle's edge. If you move the center of the circle down from the navel to the

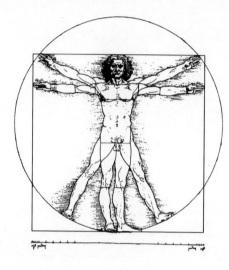

Figure 9.35. "Proportions of the Human Body"

base of the spine, you get the image of the phi ratio (fig. 9.36). The phi-ratio image in this case would occur when the perimeter of the square and the circumference of the circle are equal (fig. 9.36).

Thus, you can put a square around a body with a north-south pole running down the middle, and from that you can mathematically derive the phi ratio. See figure 9.34.

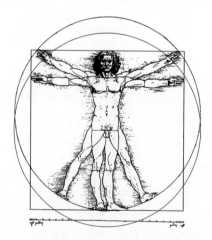

Figure 9.36. The phi ratio image.

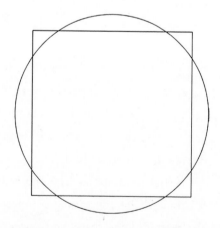

Figure 9.37. When the perimeter of the square and the circumference of the circle are equal, the phi ratio image is produced.

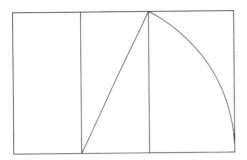

Figure 9.38. The golden mean rectangle.

The Spiral

Go back to the square that fits around the body with the line running down the middle and the diagonal. Use a compass to rotate the diagonal line and complete the rectangle by extending the two remaining lines until they meet. You then have a golden mean rectangle (fig. 9.38).

The golden mean rectangle is such that if you take its shortest edge and make a square, what is left is another rectangle proportional to the bigger one by 1.618; it goes in forever and it goes out forever. This creates a spiral that goes in and out infinitely (fig. 9.39). So the spiral is

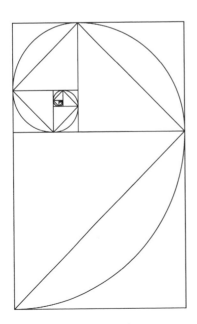

Figure 9.39. Spiral—golden mean rectangle.

derived from the golden mean rectangle. The golden mean rectangle has two fields: male energy as the diagonal of the squares, and female energy as the curved line of the spiral.

The Fibonacci Sequence

Leonardo Fibonacci, a medieval mathematician, noticed a particular order or sequence that plant life utilizes to grow and discovered that this particular ratio kept coming up everywhere. The sequence is: 1, 1, 2, 3, 5, 8, 13, 21, 34, 55, 89, 144, 233, etc. I referred to it earlier when discussing the growth of a plant.

The reason this pattern shows up consistently in life originates in the golden mean spiral, which goes in and out forever, without beginning or end. Life doesn't know how to deal with something that has no beginning because there is nowhere to start. So this sequence, which has become known as the Fibonacci sequence, is life's solution to that problem.

If you divide one term of this sequence into the next one and keep going, you will find that you quickly approximate the transcendental number 1.6180339.

For example:

1 divided by 1 = 1
2 divided by 1 = 2
3 divided by 2 = 1.5
5 divided by 3 = 1.66
8 divided by 5 = 1.60
13 divided by 8 = 1.625
21 divided by 13 = 1.615
34 divided by 21 = 1.619
55 divided by 34 = 1.617
89 divided by 55 = 1.6181

From this you can see that you keep going from under to over the transcendental number 1.6180339 sequentially. You keep getting closer and closer to the exact phi ratio of 1.6180339 without ever actually attaining it. However, you very quickly get so close that you cannot tell the difference. This is life's way of dealing with something that has no beginning and no end.

Figure 9.40 shows how it works geometrically. Use the diagonal of the first square as your measuring unit; move one unit; then make a ninety-degree turn and move one more unit; then turn ninety degrees and move two diagonals; then turn another ninety degrees and move three diagonals; then ninety degrees and five, then ninety degrees and eight, etc. . . . The spiral is unfolding exactly as in nature.

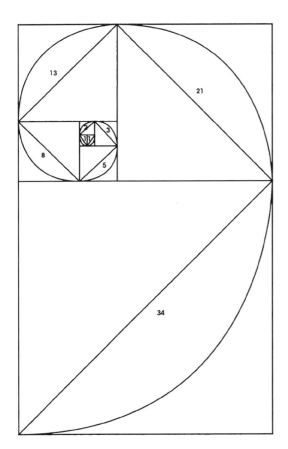

Figure 9.40.
Fibonacci spiral.

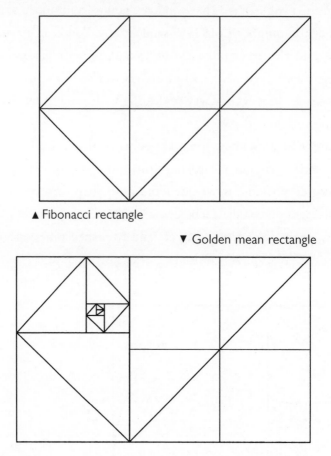

▲ Fibonacci rectangle

▼ Golden mean rectangle

Figure 9.41. Golden mean rectangle versus Fibonacci rectangle.

Figure 9.41 shows the geometrical comparison between a logarithmic golden mean rectangle on the bottom and a Fibonacci rectangle on the top. A Fibonacci rectangle is composed of six equal squares. It also has a definite beginning as compared to the logarithmic golden mean rectangle, which goes in forever. As you can see, they approximate each other very quickly.

Going back to da Vinci's sketch, we note that he drew lines on the body—on the arms in various places, on the knees, in the center, and in the chest, neck, etc. If you extend these lines you will create an eight-by-eight grid or sixty-four squares (fig. 9.42).

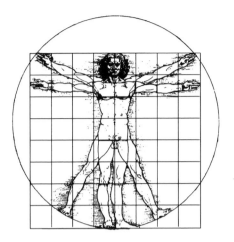

Figure 9.42.
Eight-by-eight grid.

The eight spirals of energy around the human body are based on the Fibonacci sequence. These spirals of energy come in and focus on the eight squares that surround the four central squares. See figure 9.43.

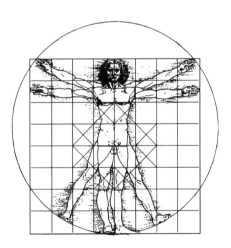

Figure 9.43. The spirals of energy focus in the eight squares
that surround the four center squares.

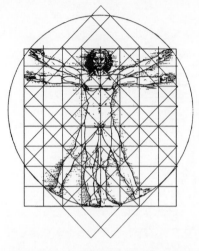

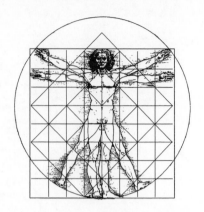

Figure 9.45. *"White light spirals"—*
they are male, and their primary
nature is electrical.

Figure 9.44. *The spirals of energy.*

Figure 9.44 shows the sixty-four squares with the spirals of energy. The spirals come in two different ways. Figure 9.45 shows one way. The starting points are the eight squares that surround the four center squares. You can trace any of these spirals using the Fibonacci sequence of 1, 1, 2, 3, 5, 8, 13, etc. These are called "white light spirals." They are male, and their primary nature is electrical.

The spirals can also go the other way, as shown in figure 9.46. If you form them this way, you have to go through the center zero point;

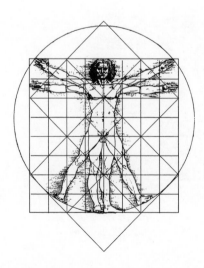

Figure 9.46. *"Black light*
spirals"—they are female,
and their primary nature is
magnetic.

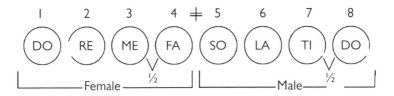

Figure 9.47. The harmonics of music.

this is the womb or void. These are called "black light spirals." They are female, and their primary nature is magnetic.

You could also superimpose the original eight cells of the human zygote, or "the egg of life," over this grid. This geometry is true for us from conception through adulthood.

The Chakra Systems

The harmonics of music and our body's chakra system are related in the geometrical pattern called "the egg of life." As shown in figure 9.47, between the third and fourth, and the seventh and eighth notes of the musical scale, there are half steps. No one seems to know why. Some people also describe a break or change between the fourth and fifth notes. There is a reflection pattern. One, two, three, half step, four; one, two, three, half step, four—there are really two sets of four. One is female and one is male.

The reason there are half steps between the third and fourth notes and the seventh and eighth notes, and a break between the fourth and fifth notes, is that sound can be explained in terms of the egg of life (fig. 9.48). As sound comes in from below it hits sphere number one. From there it has three other places to go on this tetrahedron—from sphere number one to number two, then to number three. It moves in a triangle—a flat plane in the same direction. Then, in order for the sound wave to move to the fourth sphere, it has to change direction. The fourth sphere is directly behind sphere number five. Because the sound wave changes direction it is perceived as traveling a shorter

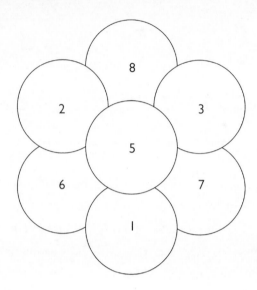

Figure 9.48. How sound is related to the egg of life.

distance, hence the half step, just as the shadow of a line appears to shorten as it changes direction. The sound wave has now completed the first tetrahedron and goes into the second tetrahedron. To do this it must go through the void in the center of the egg of life, the Great Void, to get to sphere number five. Sound changes polarity when it moves to the second tetrahedron, from male to female or from female to male. It then moves to spheres number six and seven in a flat plane, where it must make another half step to go to sphere number eight.

The eight-point chakra system resembles the movement of the eight notes on the musical scale; however, in the chakra system of the human body "the egg of life" pattern is unfolded. The chakras start at the base of the spine and move up over the head (fig. 9.49).

This is a Hindu or Tibetan system and is greatly simplified. We also have chakra systems above and below our bodies. The one below our feet is the consciousness level from which we came, and the one above our head is the next level, the consciousness toward which we are moving. They are in phi ratio. The one below is very short and the one above is very long.

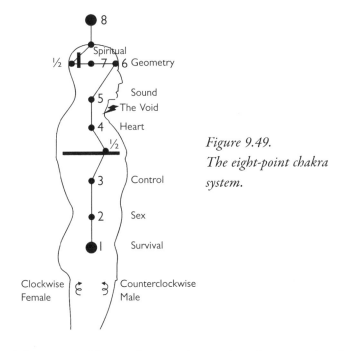

Figure 9.49.
The eight-point chakra
system.

The same half steps that we observed in the musical scale also recur in the chakra system. The chakras are like lenses though which we interpret reality. For example, when a new spirit is born its total concern is with survival—being able to stay here on this third-dimensional level. The next thing the spirit wants to do is make physical contact with other beings. Contact is first made with the mother; later it becomes sexual. Once you have established yourself and made sexual contact, establishing control follows. These represent the first three chakras from the bottom up; then there is a big wall and a half-step change of direction. You can't get through this wall until you have mastered these three chakras. Once you do get through you are at the heart, which is the fourth chakra in this system. The fifth is located at the throat and is related to music; the sixth, between the eyes, is related to geometry; and the seventh, located at the pineal gland, is the "third eye." At this point, there is another wall with another half-step change of direction. This takes us to the eighth chakra, which is above the head and points to the next phase in our conscious evolution.

This eight-point system represents only the white notes of the musical scale and, as I mentioned, it is a greatly simplified system. There are five black notes, or sharps and flats, to the musical scale as well. So there are really twelve chakra points, with the one above the head being the thirteenth (fig. 9.50). The twelve represent five sub-chakras at each site, so there are actually sixty points. Each of the twelve groups of five is separated by ninety degrees.

A straight tube runs through the body, perfectly straight like a fluorescent bulb, starting at the perineum at the base of the spine and continuing up through the soft spot in the head. The twelve chakra points fall along this line with an average of 7.23 centimeters between each of

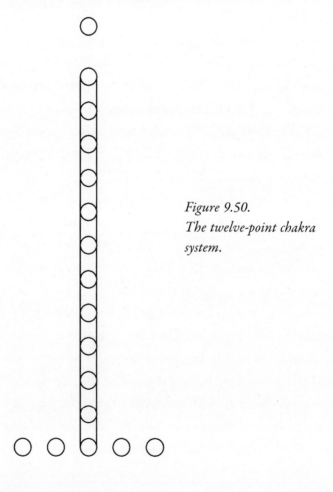

Figure 9.50.
The twelve-point chakra
system.

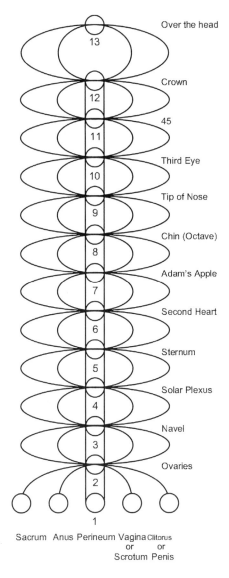

Over the head
Crown
45
Third Eye
Tip of Nose
Chin (Octave)
Adam's Apple
Second Heart
Sternum
Solar Plexus
Navel
Ovaries

Sacrum Anus Perineum Vagina Clitorus
or   or
Scrotum Penis

Figure 9.51.

the twelve points, or the distance across the palm of the average hand,
likewise the distance from the tip of your chin to the tip of your nose.

Energy spirals up through the chakra system making 90-degree
turns as it moves from one point to the next. At the base chakra
(fig. 9.51) all 5 channels are pointing toward the front in a straight

line. The opening to the vagina is a vesica piscis, and the small opening in the penis is also a vesica piscis. All the energy in these 5 points
flows front to back. As the energy moves up 7.23 centimeters to the
second chakra and the ovaries, it shifts direction by 90 degrees. Up
another 7.23 centimeters (with another 90-degree shift) is the navel.
This is where the umbilical cord was attached. The energy here moves
from back to front, the reverse of the base chakra. When we move up
again to the solar plexus, which is another vesica piscis, the energy
radiates out from side to side again, as in the ovaries. The next level
is the sternum, which is a special point affiliated with the circle
(fig. 9.52). The fifth chakra is the first instance of return; it is special
because it contains all the previous motions at once. The energy has
made a complete 360-degree rotation and knows all the directions.
Hence you have the breasts coming out to the front but also being
separated sideways. This is the Christ-consciousness point. It is at the
19.5-degree latitude of the body, and it forms a cross. At the sixth point

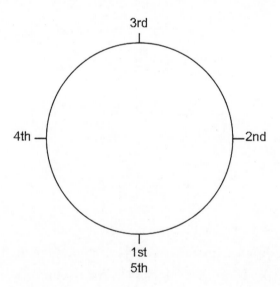

*Figure 9.52. The fifth chakra—the sternum. The energy has made
a complete 360-degree rotation and knows all the directions.*

you have the heart; at the seventh is the Adam's apple; the eighth is the chin.

Then you hit another octave and the energy runs through the head. The physical features of the face that correspond to the chakra points in the head begin with the chin. From the chin the energy rotates 90 degrees to the mouth (energy running side to side), then to the nose (energy moving back to front), then to the eyes (side-to-side energy), and then to the third eye where the energy has again made a 360-degree rotation.

The External Chakra Points

We have external chakra points too—they are points in the space around each body. We are surrounded by shape in the form of a star tetrahedron with eight external chakra points on it (fig. 9.53). If you were to photograph the internal points, they would be identical to the external ones. All the external points pulse in unison just like the internal ones. We thus have both an internal and an external aspect to our chakra system.

Figure 9.53. Our external chakra points.

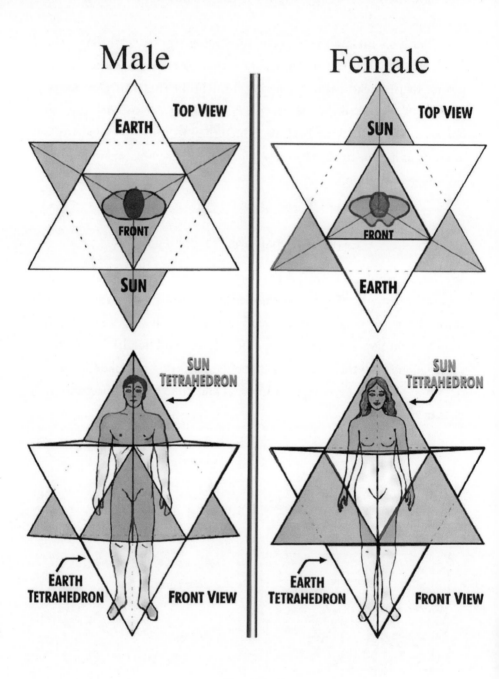

Figure 9.54.

In the components of the star tetrahedral field around each body, the one pointing up is male and is called the Sun tetrahedron. The one pointing down is female and is called the Earth tetrahedron (fig. 9.54). This is true for both men and women. There are only two ways that a person can symmetrically fit into the star tetrahedral field. If the point at the base of the Sun tetrahedron is forward, the male fits, and if the point at the base of the Earth tetrahedron is forward, the female fits. The star tetrahedron is linked to the center of the body at the base of the spine. If you jump, the external star tetrahedron goes up with you, and if you sit, it lowers with you.

The Eye

Drunvalo has drawn the morphogenic structure of the eye, be it human or the eye of any other creature. He considers it to be his most important drawing (fig. 9.55). The ocular structure is the same as the

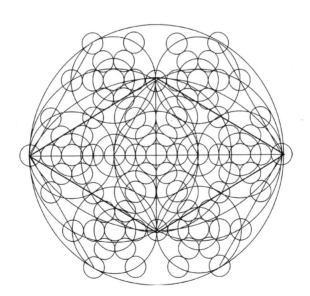

Figure 9.55.
The morphogenic structure of all eyes.

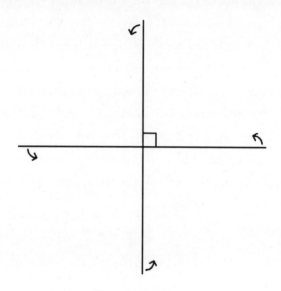

Figure 9.56.
An electromagnetic field.

structure of light itself; it contains the entire electromagnetic structure. This structure also contains the geometry of the vesica piscis. Within the vesica piscis are two equilateral triangles. The concurrent base of the triangles is the width of the vesica piscis, and the line running through their center is its length. When you rotate the length of the first vesica piscis ninety degrees, it becomes the width of the next larger vesica piscis. If you rotate this vesica piscis another ninety degrees, its length will become the width of the next larger one, and so on. This goes in forever and it goes out forever.

An electromagnetic field, or light, is an electric field with a magnetic field moving at ninety degrees to it (fig. 9.56). The electrical field is moving in a wave, and the magnetic field is moving at ninety degrees to this wave, and the whole configuration is rotating as it moves through space (fig. 9.57).

Drunvalo predicts that scientists will discover, in any electromagnetic field that the electrical aspect is the length of the first vesica piscis,

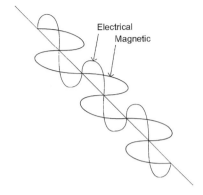

Figure 9.57.
An electromagnetic field.

and the magnetic aspect is the width of the first vesica piscis, and these are in proportion to each other. They rotate by ninety degrees as they spiral out, just like the vesica piscis. Logarithmic spirals also move along electromagnetic spirals of energy. The light field itself and the eye that receives it are in the same geometric pattern because the receiver has to tune to that which it is receiving. Our entire body is tuned to sound, vibration, music, and light.

10
The Left Eye of Horus

The Left Eye of Horus was a twelve-year "emotional body" training for aspiring Egyptian initiates. It dealt with various emotions, feelings, fears, and both positive and negative aspects of the chakras. All the temples in Egypt were built for Left Eye of Horus trainings. In this course, initiates worked with many different aspects of human nature. There is a specific fear related to each of the chakras, and each of the twelve major temples dealt with the fear related to a specific chakra.

The temple at Kom Ombo is unusual because it is the only temple dedicated to two gods. The southern half was dedicated to the crocodile god Sobek; and the northern part of the temple was dedicated to Horus. In this temple there is a huge hole in the ground (fig. 10.1). Even modern Egyptian guides tell tourists it was part of a training process: an initiate had to go into this hole, which was filled with water. Huge stones were placed inside, so they had to be careful not to hit them. The training included getting to the bottom, passing through a small opening, and then coming out the other side.

It doesn't seem so difficult, and it isn't. But this is only a partial picture of what the initiates were doing here. Egyptologists know from ancient writings only that there was water in the hole and that it was

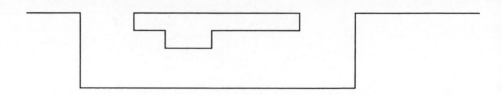

Figure 10.1. The hole in the ground.

part of a training, but they really do not know the particulars. Drunvalo asked Thoth for an explanation as to the real initiation, and Thoth gave him the following account:

The temple complex was situated such that the initiation began on a much higher level. The students could see only three steps leading into water at the beginning, and then a very high wall on the other side (fig. 10.2). The initiates had to go down the three steps and into the water, and then in one breath come out but not the same way. They had to work their way slowly because it was dark; it was thirty feet or so to the bottom, and there were obstacles. So they carefully made their way down to the bottom of the hole, where they found a passageway leading to a huge tank. This tank was lit up and they could see that they had company; it was filled with crocodiles!

Now suppose for a moment that it is you in the tank. How would you react? Would you remain calm and centered as you searched for the other exit; or would you panic and get out of there as fast as you could? Well, if you reacted as almost all of the initiates did, you would quickly weave your way through the crocodiles and head for a clearly visible opening at the top of the tank. Great you made it—but did you?

The whole purpose of this training was to give the student an opportunity to integrate his or her fears into high levels of alertness—this is what happens to either a perceived or real threat to your safety when it integrates. Doing so would have given the initiate enough presence to search for and find the correct exit. Taking the obvious way out was a clear indication that the person was in a contracted state and still at the effect of his or her fears. In other words, they failed.

This meant that after much more training, and then only if it was felt that they were ready, they had to try again. The second time the initiates knew two things; they knew the opening they had used was not the correct way out, and they knew that the tank was filled with crocodiles. This could easily have made the adventure far more fearsome than it ever was the first time. They now knew that they would soon be given an involuntary opportunity to stay present in the midst of their greatest fears! Doing so enabled enough of the fear to be transmuted into alertness, thus enabling them to see that by going deeper into the tank, that there was indeed another exit, the correct one.

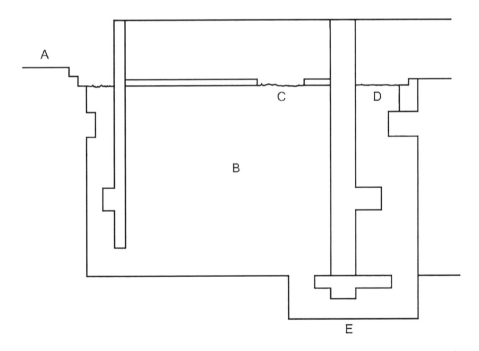

Figure 10.2. Left Eye of Horus training. A: the initiate begins here. B: tank filled with crocodiles. C: the incorrect exit. D: the correct exit. E: the Figure 10–1 hole in the ground.

By the way, the crocodiles were well fed; they weren't trying to do anybody in. They only wanted to give them an experience that they probably otherwise would never have chosen to have. Such was life for the initiates in the ancient mystery schools!

The Great Pyramid

After twelve years of the Left Eye of Horus training (the emotional body training) and twelve more years of the Right Eye of Horus training (the Unity consciousness training), the Egyptian student would descend into the Great Pyramid for a three-and-a-half-day period of final initiation.

Thoth says that the pyramids were built specifically to bring someone from the second level of consciousness (where we are presently) into the third level, which is Christ-consciousness. The Great Pyramid was an initiation chamber.

The King's and Queen's chambers were given these names by the Muslims because of their suggestive design. The roof of the King's chamber is flat, and Muslims buried their men under flat roofs. The Queen's chamber has a pitched roof, and the Muslims buried their women under similar roofs. According to Drunvalo, these chambers had nothing to do with burials—they were initiation chambers, period.

The initiation didn't begin inside the Great Pyramid. It began underneath it. It then moved to the King's chamber and finally to the Queen's chamber (fig. 10.3). The process began in a room under the Great Pyramid, enabling the initiates to encounter a "black light" spiral that goes into the center of the Earth as well as into the Halls of Amenti. Such a tunnel still exists and seems to end in the middle of nowhere. No one today seems to know why.

According to Drunvalo, the tunnel is located where it is in order to connect to the "black light spiral" at the soonest possible point after it has passed through zero point or the Great Void. This tunnel is in fact a "black light spiral" initiation chamber. Whatever you think in that tunnel becomes real because it is a true fourth-dimensional space. Many people have died here because they manifested their fears. They probably got tired of hauling dead bodies out of there, so about forty years ago the Egyptian government blocked it off from tourists.

From there the initiation moved to the King's chamber, a space designed to catch the "white light spiral" at its source, and to filter

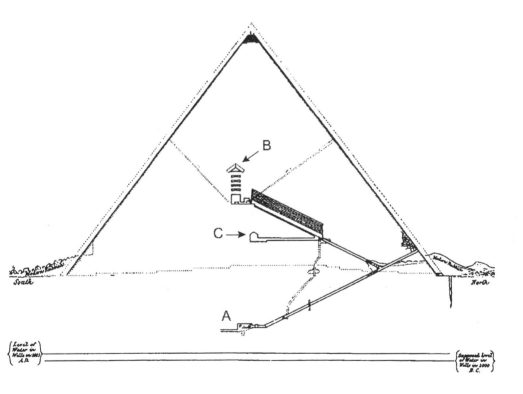

Figure 10.3. Initiation in the Great Pyramid. The initiation began underneath the Great Pyramid (A). Then it moved to what is today called the King's chamber (B) and finally to the Queen's chamber (C).

out the black light one. The King's chamber seems to be off center, but the sarcophagus was placed in such a way that, if you were lying in it, the white light spiral would go directly through your pineal gland. The initiates would recline in the sarcophagus for three and a half days, leave the third-dimensional world, and experience incredible consciousness expansion. They could then find their way back to their bodies because they were following the Fibonacci spiral and not the golden-mean logarithmic spiral. As I discussed earlier, the logarithmic spiral has no beginning and no end, but the Fibonacci spiral does have a definite beginning. This enabled initiates to trace their way back into their bodies.

After this mind-boggling transformation of self and reality, the student then went to the Queen's chamber, which served as a stabilizing room. Once a person had gone through the experience in the King's chamber successfully, he or she, needless to say, was altered dramatically and needed some settling. This is what the Queen's chamber was—a sanctuary for stabilizing Christ-consciousness.

When archaeologists first opened the sarcophagus in the King's chamber, they found an unusual white crystalline powder that they scooped up to examine. The powder is now in the British Museum. No one knew what it was until recently, and the explanation is the last thing anyone expected. Scientists have since found that when you are in a particular deep state of meditation, you excrete a certain chemical from your pituitary gland, which crystallizes into a powder. There was quite a lot of this powder found in the sarcophagus in the King's chamber, indicating that there must have been many people initiated there.

Many people have theorized that the Great Pyramid was a burial place. However, there is a lot of evidence to suggest that it was not a burial place at all but was in fact a place of initiation. Without exception, in every case in ancient Egypt that a person of prominence died, the priests cut out the heart and various organs and placed them into four jars. They mummified the person, put the body into a sarcophagus, placed the lid on, and sealed it. The body was then carried to a burial place. The sarcophagus in the King's chamber is bigger than the doorway. This means it had to have been put there while the structure was being built—not the custom for burials at all.

11
The Hall of Records

According to Edgar Cayce, the famous psychic, the opening to the Hall of Records, which holds the history of the Earth, will be found in the right paw of the Sphinx. This has been clearly marked geometrically. Looking at figure 11.1, if you bisect the golden mean rectangle that fits around the spiral at the Giza plateau, it passes exactly through the headdress of the Sphinx. Also, a line extended from the southern face of the middle pyramid and the line that bisects the golden mean rectangle form a cross that marks a very specific spot on the right shoulder of the Sphinx.

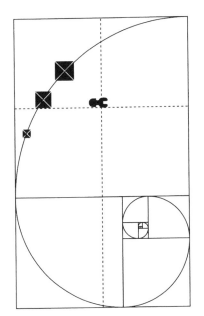

Figure 11.1.
The opening to the Hall
of Records is geometrically
marked in the right shoulder
of the Sphinx.

The Sphinx has, until recently, been under major renovation. One of the problems is that the right shoulder—the area marked by the cross—keeps breaking open. A seven-hundred-pound chunk of it fell off in 1988, and the head is also falling off. Thoth said that the head will fall off and that, in the neck, there will be found a large golden sphere, which is a time capsule. The Egyptians are doing everything they can to keep the head and right shoulder together.

Thoth says everything was set up at a higher level so that the Hall would be discovered before the end of 1990. He said that a hundred and forty-eight sets of three people would try to enter, until one of these sets, coming from the West, would open the doorway by making a sound with their voices. Inside there would be a spiral staircase going into an underground room. The Japanese have the technical ability to see this room clearly enough to detect a clay pot in the corner.

There are three channels that go out from this room. If you know how to read it, the clay pot will tell you where to go and what to do. Thoth said that the three people from the West would enter and go down the right channel. If you go down the wrong channel, or if you are not the right group, you will die—a real "Indiana Jones"-type scene. Of course, we all know that Indy would have found a way.

If you are the chosen group, you can walk right in without any problem. These three eventually will come down a long stone hallway lit on its own with no lights; that is, the air itself would be luminous. High up on the left side of the wall would be etched forty-eight sacred geometry drawings. These are the illustrations of the chromosomes of Christ-consciousness, the first one being the flower of life. At the end of the hallway there is a slight right-hand turn into a large room. Sitting on the raised shelves of this room is physical evidence of the existence of civilization on this planet for the last 5.5 million years.

At the front of the room is a stone. At the top of the stone these three people would find something like a photograph, an image of themselves. Beneath the images in the photograph, they would find their names—not necessarily the names they were given at birth, but their true names. Underneath the names would be a date, which would be that actual day. Originally Thoth was going to meet them, but he is now gone. Unless it happened before May 4, 1991, someone else will meet them. He said that each of these three people will be allowed to remove one of these objects and take it out.

The Hall of Records contains more than physical objects. Information is stored on many different dimensional levels.

12
Prana

When the poles last shifted in 10,500 BCE and we fell in consciousness, we stopped breathing in the ancient way. The way we have breathed since is very unusual. Hardly anyone else in our universe does it this way.

In general, there are two things we take in when we breathe. There is air and there is prana. Prana is life-force energy itself, more vital than air for our existence. Prana is not just in the air, it is everywhere—it even exists in a vacuum or a void.

Prana exists as an energy field connected so closely with spirit that spirit cannot exist without it. If you take air away you have a couple of minutes before you die; if you take water away you have even more time; and if you take food away you have more time still; but if you break prana from spirit, death is instantaneous. So, taking in prana with breath is a crucial act in sustaining our form of life.

We are supposed to breathe such that while air comes in through our mouth and nose, we take the prana in through the top of our head—that is, what once was the soft spot on the top of our head. Simultaneously, we take prana in from below through the perineum. The prana channel through the body is about two inches in diameter and extends one hand length above the head and one hand length below the feet. It connects with the crystalline grid around the body.

The prana then comes from above and below the body and meets in one of the chakras. The chakra in which it meets depends on where you are mentally, emotionally, and dimensionally "tuned." This is a very specific science.

After the last pole shift thirteen thousand years ago, we stopped breathing in this manner and started taking in the prana through our mouth and nose directly with the air, thus bypassing the pineal gland in the center of the head. The pineal gland is an eye, the third eye, not the pituitary gland. It is shaped like an eyeball, round, hollow, and with a lens for focusing light and color receptors. It is designed to receive light from above to go to every cell in the body instantaneously. As it is shaped like an eye it is also a *vesica piscis,* the image that emerges on the first day of Genesis. This geometrical shape is the beginning of all creation, and contained within it is all sacred information of the universe.

When this gland is not activated, however, it turns off. Normally it should be about the size of a quarter, but in us, it has become the size of a pea because we haven't used it for about thirteen thousand years. The direct result of turning off the pineal gland is polarity consciousness— good and bad, right and wrong.

Because of the way we breathe, we see things in terms of good and evil, but in fact there is no such thing as polarity. Everything actually has three components—the holy trinity—and no matter what polarity you think of there will be a third element to it. For example, with hot and cold, there is warm; up and down gives you middle, etc. From higher-dimensional levels of existence polarity is just an illusion. Unity is all there is; there is just one God and One Spirit that moves through everything. All that ever occurs then is occurring because of the one God.

On our level of existence, we interpret things differently. We think we see good and evil. What is really going on, however, is timing. The forces of polarity are necessary for the proper functioning of the universe. Remember, the dark forces do everything they can to hinder a particular area of consciousness, be it on a planetary or individual scale. The light forces do everything they can to encourage awareness. This opposition causes consciousness to move upward at exactly the right pace. In the birth of a human child, for example, nine months is the proper gestation time, not three months or fifteen months. The forces of polarity cause the child to be born at exactly the right time.

So given where we are, we need to see good and evil and be aware of them, but we also need to recognize that the presence of God is in every situation, and there is a reason for everything that happens. We need to see that everything is whole and complete and perfect, and that there is nothing wrong no matter how bad or how good it may seem. We need to see that life engenders a deep conscious aspect that is everywhere.

In our state we tend to think that we live inside this body, that everything "out there" is separate from us, and that our feelings and thoughts don't manifest beyond ourselves. We think that we can hide our thoughts and feelings and they have no impact "out there." This is simply not true. Everything that we think, feel, and do is creating our whole world, all the way out to the most distant stars. We are creating everything in every moment, more so than we could ever imagine.

Christ-Consciousness Spherical Breathing

Drunvalo brought forth a meditation, given to him by his angels, designed to get us back to breathing in the ancient way. It involves taking prana in through the top of the head and also through the perineum so that energy comes from above and below the body. The prana then connects with the crystalline grid around the body (the star tetrahedron inscribed in a sphere) and meets in one of the chakras.

The meditation is done in a series of fourteen breaths. The first six breaths are for balancing the polarities within the eight electrical circuits and also for cleansing these circuits. The next seven breaths reestablish the proper pranic flow through the body and recreate spherical breathing within the body. The fourteenth breath changes the balance of pranic energy within the body from third-dimensional to fourth-dimensional awareness.

This meditation is much more than just breathing, however. It uses a combination of mind, body, breath, and heart, all working together in harmony.

These breaths are the first fourteen in a series of seventeen used to create the counter-rotating fields of the Merkaba around the body. Because the internal Merkaba can only be created with the emotional body intact, it is absolutely necessary to open your heart and feel love and unity for all life while doing this meditation. You do this to the best of your ability, of course.

You'll be utilizing hand positions called *mudras* that connect you to a specific electrical circuit within your body. You will be changing mudras in each of the first six breaths, thus connecting you to different electrical circuits. There are eight electrical circuits in the body, coming from the eight original cells. It is necessary to balance only six circuits; in the process of doing that, the one above the head and the one below the feet will automatically be balanced.

You begin by sitting in a comfortable, relaxed position with your spine straight. Close your eyes and let the outside world go. When you feel calm and relaxed, expand your feelings to a state of love and unity for all life everywhere, and also visualize the star tetrahedron around your body.

It is necessary to know clearly your relationship to the star tetrahedron as you are standing or sitting, see figure 9.54 (p. 140). The apex of the Sun tetrahedron is always one hand length over your head, whether you are standing or sitting. Remember, if you are standing, the base of the

Sun tetrahedron is just above your knees. If you sit down, the base moves down accordingly. That means, then, if you are sitting in a chair, the base is on the floor, or at least very close to that. If you are sitting on the ground, approximately one-half of the Sun tetrahedron is in the ground.

The base of the Earth tetrahedron is always at the solar plexus, whether you are standing or sitting. If you are standing, the apex of the Earth tetrahedron is one hand length into the ground. If you are sitting on a chair, the apex moves down accordingly. If you are sitting on the ground, more than one half of the Earth tetrahedron is in the ground. The length of all sides of the tetrahedrons is equal to your height; so if you are six feet tall, the length of all sides of both the Sun and Earth tetrahedrons is also six feet.

On the inhale of the first breath, visualize the Sun tetrahedron. This is the one with the apex pointing up. The point at the base of the tetrahedron, which is just above the knees when you are standing and, on the floor, if you are sitting, is facing toward the front for males and toward the back for females. Visualize to the best of your ability this Sun tetrahedron filled with brilliant white light, the color of lightning. Your body is surrounded by this light. If you cannot visualize it then "sense" or "feel" it around you.

Also, on this first breath arrange your hands, palms up, with the thumb and index finger lightly touching. This is a mudra. Do not allow any of your other fingers to touch one another.

Inhale through your nose in a deep, relaxed, rhythmic manner for seven seconds, bringing the breath up from the stomach, then to the diaphragm, and finally to the chest. Do this all-in-one movement.

Then, without pausing at the top of the inhale, begin your exhale. Exhale slowly through your nose for seven seconds. As you exhale, visualize the Earth tetrahedron. This is the one with the apex facing down, the point at the level of the solar plexus facing the back for males, and the front for females. Again, visualize this tetrahedron filled with brilliant white light, the color of lightning.

After you have completed the exhale in seven seconds, relax and hold your breath for approximately five seconds. Move your eyes toward each other (look slightly cross-eyed), then look up and immediately look down to the ground as fast as you can. At the same time visualize the white light in the Earth tetrahedron shooting out through the apex of the tetrahedron and into the center of the Earth.

As you are doing this, you should feel an electrical sensation moving down your spine. This is called "pulsing." What you are doing is clearing out the negativity in the part of your electrical system that is associated with the mudra you used (index fingers and thumbs touching).

Immediately after pulsing the energy down your spine, begin the second breath. The second breath is exactly the same as the first breath except for one thing. The one and only change is that a different mudra is used. For the second breath you hold the thumbs and second fingers together. Everything else is the same as the first breath. Similarly, for breaths three through six, everything in these breaths is the same as the first breath except for the mudras.

For the third-breath mudra, you hold the thumb and third finger (ring finger) together. For the fourth-breath mudra, hold the thumb and little finger together. For the fifth-breath mudra, hold the thumb and index finger together just as in the first breath. And for the sixth-breath mudra, hold the thumb and second finger together, just as in the second breath.

The next seven breaths begin a different breathing pattern. It is no longer necessary to visualize the Sun tetrahedron on the inhale and the Earth tetrahedron on the exhale. What you visualize instead is the tube that runs through the body. This tube extends one hand length above the head and one hand length below the feet. In other words, the tube runs through the apex of the Sun tetrahedron, which extends one hand length above your head. It also runs through the apex of the Earth tetrahedron, extending one hand length below your feet. The diameter of your tube is exactly the same as the diameter of the hole formed when your thumb and middle finger are touching.

Begin the seventh inhale immediately after the pulse following the sixth exhale. Inhale rhythmically, taking about seven seconds, just as you inhaled for the first six breaths. As you begin the seventh inhale, visualize the tube running through your body as well as brilliant white light running up and down the tube at the same time. In other words, visualize prana running down the tube from over your head and simultaneously running up the tube from beneath your feet.

Now visualize the light meeting inside the tube at the level of the navel or third chakra. As the two beams of light or prana meet, a sphere of light or prana about the size of a grapefruit forms and slowly begins to grow.

This all happens in the instant you begin the seventh inhale. As you continue to inhale for seven seconds, the sphere of prana slowly grows. At the end of the seventh inhale, immediately begin your exhale. There is no more holding of the breath and no more pulsing.

For the next seven breaths use the same mudra, that is, both the index and second fingers lightly touching the thumb with the palms up.

As you begin to exhale, the prana continues flowing from each end of the tube and expanding the sphere centered at the navel. By the time of the full exhale (seven seconds), the sphere of prana will be about eight or nine inches in diameter.

Begin the eighth breath immediately after the seventh exhale. On the eighth breath the prana sphere continues to grow until it reaches its maximum size at the end of the exhale. At its maximum the sphere has a radius of one hand length.

On the ninth breath, the sphere cannot grow bigger, so what it does is grow brighter. Visualize the sphere growing brighter and brighter on both the inhale and exhale.

On the tenth breath, continue to visualize the sphere growing brighter. About halfway through the inhale, the sphere will reach critical mass and ignite into a sun. As you begin to exhale, make a small hole with your lips and blow the air out your mouth like you are forcing it out. Then let it all go with a final *whoosh*. As you do, the ignited sun expands outward to form a sphere around your body. This is the same sphere shown in Leonardo da Vinci's drawing. Your whole body is now inside a sphere of charged white light or prana.

The sphere is not yet stable at this point, however. It took all your energy just to get it out there. It will take three more breaths to stabilize it—the eleventh, twelfth, and thirteenth breaths. Inhale and exhale just as you did for the seventh through ninth breaths, all the while feeling the flow of prana through the tube, meeting at the navel, and expanding into the sphere around your body.

The sphere is now stabilized and you are ready for the all-important fourteenth breath. For the seventh through the thirteenth breaths, the prana flow met in the tube behind the navel. That tuned us to our third-dimensional reality. If we were going to stay here we would stop after thirteen breaths. Since we are moving to the fourth dimension, the fourteenth breath becomes necessary in order to retune us to that reality.

At the beginning of the inhale of the fourteenth breath, you move the point where the two streams of prana meet up from the navel to the sternum. The entire large sphere around your body moves up as the original small sphere, which is also still contained within the large sphere, rises to the sternum (fig. 12.1). Having the prana meet here tunes you to fourth-dimensional or Christ-consciousness.

As you do this you change the mudra also. Males place the left palm on top of the right palm with the thumbs lightly touching; females place the right palm on top of the left palm with the thumbs lightly touching. Keep this mudra for the remainder of the meditation.

As you continue to breathe from your Christ-consciousness center, switch to shallow, relaxed breathing and let yourself feel the flow of prana and love for as long as you like.

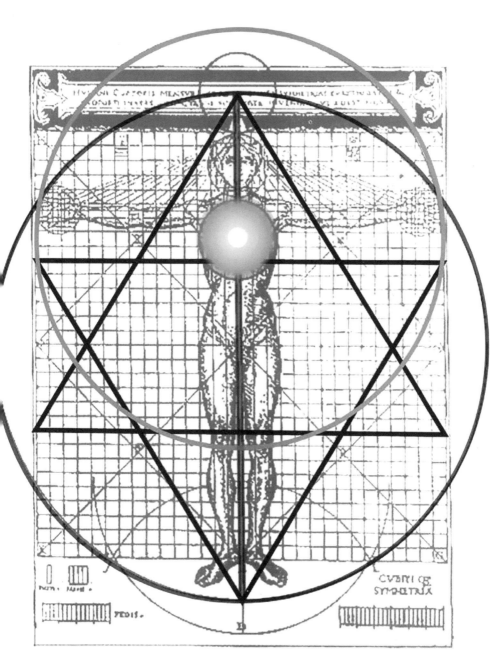

Figure 12.1.
Vitruvius's Canon, showing the prana spheres centered at the sternum.
The star tetrahedral field and the prana tube are also shown.

Activating the Light Body

Breaths fifteen through seventeen are the ones that set the counter-rotating fields of the Merkaba in motion.

As you prepare for the fifteenth breath, keeping the same mudra, be aware of the whole star tetrahedron; and remember, there are three identical star tetrahedral fields superimposed over each other.

The first star tetrahedron is neutral in nature and is literally the physical body. It is locked in place at the base of the spine and does not rotate. Like the other two star tetrahedral fields, it is placed around the body according to maleness or femaleness.

The second star tetrahedron is electrical in nature, male, and rotates to the left. It is literally the mental body.

The third star tetrahedron is magnetic in nature, female, and rotates to the right. It is the emotional body.

On the seven-second inhale of the fifteenth breath, say to yourself, "Equal speed." This will start the two whole-star tetrahedral fields spinning in opposite directions at the same speed. This means that for every complete rotation to the left of the mental (male) star tetrahedron, there will be a complete rotation to the right of the emotional (female) star tetrahedron.

You exhale on breaths fifteen through seventeen in the same "forced air" manner as you did for breath number ten. As you exhale, the two sets of star tetrahedrons start spinning, instantly reaching one-third the speed of light.

On the inhale for the sixteenth breath, say to yourself, "Thirty-four, twenty-one." This will start the two-star tetrahedrons spinning in opposite directions at a ratio of thirty-four spins to the left for the mental (male) star tetrahedron, and twenty-one spins to the right for the emotional (female) star tetrahedron.

On the forced-air exhale, the two sets of star tetrahedrons instantly increase in speed from one-third to two-thirds the speed of light. Also, a fifty-five-foot-diameter disk forms around the body, centered at the base of the spine. This disk, along with the sphere of energy that is centered on the two sets of star tetrahedrons, creates a shape around the body that looks much like a flying saucer. This is the Merkaba. However, at this point the field is not stable. More speed is needed, and the next breath provides that.

As you inhale on breath seventeen, say to yourself, "Nine-tenths." This will stabilize your Merkaba field by increasing the speed to nine-tenths the speed of light. This breath will also tune you perfectly to our third-dimensional universe, where electrons rotate around atoms at exactly this speed.

As mentioned above, you exhale in the same forced-air manner as you did for breaths fifteen and sixteen.

After you have completed breath seventeen, even though you could get up immediately and continue with your everyday life, it is far better if you can remain in the meditation a while longer. While you are in this meditative state, your thoughts and emotions are amplified. This is a great time for positive affirmations and the setting of intent.

There is ultimately one more breath in this meditation, the eighteenth breath. This is the breath that, when taken, will carry you through the speed of light and into the fourth dimension. This breath can only be given to you from within, from direct contact with your higher self.

If you would like some assistance in learning this meditation, you might be interested in knowing about a workshop that I gave from 1995 to 2016, called the *Flower of Life*. In this workshop, you are given information that has been kept secret for more than three thousand years. This includes everything you always wanted to know about our ancient past and our intergalactic connections. We also work with the universal language of sacred geometry, which is used to show the mind the Unity of Being. Ultimately, the workshop is about the teaching of the Merkaba and Unity Breath meditations, and the activation of the energy fields around the body.

I have also created a video version of this workshop, called *The Flower of Life Workshop: Igniting Your Inner Light*. In this twenty-hour video course, you will be given everything that I have presented in the Flower of Life workshop, including the teaching of the Merkaba and an accompanying MP3 recording that perfectly guides you through both the Merkaba and Unity Breath meditations. You can find more information on my website.

Unity Breath Meditation

Now I would like to give you a meditation that came from Paramahansa Yogananda's guru, Sri Yukteswar and is one of the final phases of initiation into Kriya yoga.

In order to do this meditation, you need to be aware of your prana tube that begins one hand length above your head and terminates one hand length below your feet. Breathe prana through your tube from above and below, and let it meet in your heart chakra (at your sternum). From there it will radiate into a sphere around your body.

You also need to be aware of the new grid around the Earth. Just know that it is there, and at one point in the meditation, you will connect with it.

Close your eyes. You don't need to know the Merkaba, but if you do, go into it, or if you know the Christ-consciousness spherical breathing go into it; otherwise, it doesn't matter.

Relax and take a deep breath, see a white mist as you exhale, and relax into it. Take one more breath and feel your body relaxing.

Place your attention on the tube that runs through your body. With your intention, allow the two ends of the tube to open, and feel the white light of prana come rushing in and meet in your heart and then radiate out into a sphere around you. The flow is continuous, whether you are inhaling or exhaling. Just feel this for a minute.

Now bring your attention to Mother Earth and to Nature and feel the love that you have for her, for the trees the clouds, the wind, the birds, the animals, the water, the people. Feel the love you have for the Earth and center it in your heart, right where the prana flows are meeting. Bring it into a little sphere.

When you get your love there where you can feel it, send it down to the center of the Earth, and wait for Mother Earth to send back her love for you, up the tube. Wait until you can feel this.

Now place your attention on Father Sky, to all life everywhere; all the stars the galaxies, all the life in the heavens. Feel your love for the Father, think of looking into a night sky.

Take all your love for Father Sky and put it into your heart, where your love for the Mother is. Keep them separate for a moment. Now send it up and let it connect with the Christ-consciousness grid, about sixty miles above the Earth. Then wait until you feel the love of Father Sky come back.

With the love of Mother Earth and Father Sky in you at the same time, the holy trinity is present—mother, father, and child. When this happens, something very special can take place.

Become aware of the tube again. This time, place your attention on the two ends. Open them even further, and allow all life everywhere, the consciousness of all creation, to enter from both poles, to come into your heart, and to radiate as light around you. Open them up, and allow God to come in, and to form a sphere around your body.

Now allow that sphere to expand, slowly at first. Then let it get bigger and bigger, moving out faster and faster. Finally let it expand uncontrollably, through all dimensions. Let it return back to all life everywhere.

Now you are breathing all life, and all life is breathing you. May your life never again be the same!

13
The Philadelphia Experiment

At least on the surface, the Philadelphia Experiment (officially known as Project Rainbow) was a top-secret military experiment that attempted to make a battleship invisible. Yes, invisible! Did you see the movie of the same name? This happened in 1943, right in the middle of World War II. Making a battleship invisible is really not that difficult. All you have to do is take it into the next overtone and it is invisible to everyone on this dimensional level; it's the cloaking device of the Klingons. The technology for the Philadelphia Experiment came to us from the Greys. We wanted it to win the war, but they had very different reasons for helping us, which I will discuss later.

To go to the next higher dimensional overtone, you have to create counter-rotating fields of energy at very specific speeds. In the experiment, the government scientists got counter-rotating fields going based on a star tetrahedron. When you go from one dimensional world to another, the counter-rotating fields shoot from nine-tenths the speed of light to the actual speed of light, which involves an incredibly complex series of whole-number harmonics that build upon each other.

The visual experience of this is that the space around you turns to a red fog and takes the shape of a flying saucer. The colors then progress through the whole rainbow very quickly, going from red to orange, yellow, green, blue, purple to ultraviolet purple, and then to a blinding white light that slowly recedes. Any physical objects will appear to be made out of gold, which will slowly become translucent and then transparent. Then you are into blackness, and at this point you make a ninety-degree shift, which is done in two distinct forty-five-degree turns. Different dimensional worlds are all separated by ninety-degree angles. After the ninety-degree shift you reappear in a whole new world on a different dimensional level.

The Philadelphia Experiment, regardless of whatever was actually done, was a true-life episode carried out by the U.S. Navy in 1943. It involved the battleship, *USS Eldridge*. The goal of the project was to make the ship invisible to radar, not totally invisible. In this experiment the colors went from red to orange to yellow to green—that much didn't take very long—but at that point the researchers panicked and tried to shut it down, so they never got beyond that stage. It was like taking a jet a few hundred feet off the ground and then turning off the motors; the experiment inter-dimensionally crashed. The ship disappeared out of the Philadelphia Navy Yard for about four hours. When it reappeared some of the crew members were found literally imbedded in the deck, two were found in bulkheads, some were on fire, some didn't come back at all, and some were repeatedly materializing and dematerializing. The survivors were all completely disoriented.

Two particular individuals jumped off in the middle of the experiment thinking they could swim away, but when they landed, they found themselves not in the Philadelphia harbor but on Long Island, New York, in 1983. The reason they landed at this time was because a similar experiment, called the Montauk Project[1] was carried out in 1983 and was connected to the 1943 Philadelphia Experiment. These two individuals were brothers named Duncan and Edward Cameron.

Both experiments were conducted on August 12th of their respective years. According to Al Bielek, who claims to be Edward Cameron, one of the two men who jumped off the *USS Eldridge*, there are four bio-fields of the planet, and all four peak out every twenty years on the 12th of August—1943, 1963, 1983, 2003, etc.[2] This creates a peak of magnetic energies at these times and also coupled the two experiments. The energies were sufficient to create a hyperspace field and caused the battleship *USS Eldridge* to slip into hyperspace during the 1943 experiment.

Bielek says the linking of the two experiments, which ripped a huge hole in space-time, was caused deliberately by the Greys who worked on the 1983 end. They did this to put a rift in the fabric of space-time so that large numbers of aliens and ships could come through. Evidently the hole was necessary to get large ships through and make a mass (silent) invasion of the United States. From that invasion came the joint U.S.–Grey treaty. In linear time, the 1983 Montauk Project came after the 1943 Philadelphia Experiment, but this experiment created a time-loop such that the way we conceive of linear time chronology no longer applies.

It was Duncan's spine that was used to run the fields in both the Philadelphia and Montauk experiments. That is just what life forms do—whales and dolphins, for example. The head of the pod uses his spinal column to set up an electromagnetic field around him, and everyone else in the pod is connected to that field. It is as though there is just one body with many separate cells in it. Whatever the lead being does, everyone else follows. This explains the phenomenon of beached whales: if the lead whale ends up on the beach for whatever reason, all the other whales in the pod will follow. They are acting as one body.

Anyway, Duncan Cameron was used in both experiments. The hole in space-time has created huge vortices of energy in the fourth dimension. If these vortices of energy appear in our dimension, huge areas of the Earth could be destroyed—even the entire planet. However, higher beings are not concerned about this; they say there is no problem.

The Monuments of Mars

It was the Greys who set up these time experiments, not our government. It wasn't even really our government who followed their lead; it was something called the "secret government," and the experiment was not done for the purpose of making a ship invisible, although that was the ostensible reason. The Greys had a much larger agenda having to do with Mars. Remember, one million years ago their ancestors were successful with a similar experiment almost identical to this one; in fact, they left the monuments on Mars mathematically describing that experiment. These monuments were first photographed by NASA's *Viking* in 1976.

The *Viking* photographs show what looks like a face on Mars in an area known as Cydonia. *Viking* also photographed a five-sided pyramid, several four-sided pyramids, and other distinctive objects.

A number of people, including Vincent di Pietro and Gregory Molenaar, NASA subcontractees, and later Richard Hoagland,[3] got the *Viking* photographs from NASA and released them. After interpreting the photographs of monuments on Mars, which NASA was very reluctant to acknowledge, Hoagland, along with geologist Erol Torun, began exploring the geometry of the objects in them. As he learned more about the science of sacred geometry and studied in great depth the images, Hoagland discovered that the angles between the pyramids described mathematically in great detail a tetrahedron inscribed in a sphere. When you have a tetrahedron inscribed in a sphere with the apex placed at either pole, the base will touch the sphere at 19.47 degrees (rounded off to 19.5 degrees) of the latitude opposite that pole.

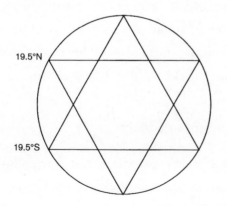

Figure 13.1.
A star tetrahedron inscribed in a sphere. The two bases touch the sphere at 19.5 degrees north and south latitude.

If you have a star tetrahedron (two back-to-back tetrahedrons) inscribed in a sphere, the two bases will touch the sphere at 19.5 degrees north and south (fig. 13.1). Hoagland found this angle repeated over and over in the images of the pyramids on Mars. This turns out, of course, to be a very significant angle.

Hoagland also studied images of Jupiter, Saturn, and Neptune taken by various unmanned NASA spacecraft. He found that the red spot on Jupiter sits at 19.5 degrees latitude. The great dark spot on Neptune is at 19.5 degrees latitude; and researchers believe there is a similar dark spot on Uranus at 19.5 degrees south latitude. On Saturn there are two cloud bands at 19.5 degrees latitude north and south. In addition, when you track sunspot activity, its primary focus seems to be 19.5 degrees north and south latitudes. Thus, there are major concentrations of planetary energy at 19.5 degrees north and south. On Earth, the Mauna Kea Volcano on the Big Island of Hawaii is found at 19.5 degrees north.[4]

According to Drunvalo, there are actually three star tetrahedral fields around all planets, and they are rotating and creating electromagnetic fields. NASA and Hoagland are at present time only becoming aware of one of these fields.

NASA did not want us to know about the discovery of pyramids on Mars and has done everything it can to ridicule interested parties, even those within their organization. Why does NASA not want us to know about this? It has to do with the secret government, the Greys, and Earth changes, which will be presented later.

Free-Energy Machines

There now exist working free-energy machines, based upon this technology of rotating energy fields. One is called the N machine, which according to its creator Bruce DePalma, can generate up to four times more power than it consumes. This, of course, defies the basic law of the conservation of energy, which says that the output of energy cannot be more than the input. And of course, most physicists simply refuse to look at DePalma's findings and dismiss his theories out of hand. These machines however can provide virtually an infinite supply of electricity for just a few hundred dollars. They never wear out.

Nikola Tesla tried to give the world free energy a long time ago. According to George C. Andrews in *Extra-Terrestrial Friends and Foes:*

> Tesla intended to offer to the world free and inexhaustible energy, via the tapping of the electricity from the earth and atmosphere, then broadcasting it through a wave carrier, just like radio. That break-through was obviously not to the taste of the tycoons, owners of electric plants, generator and electric cable manufacturers, not to mention the oil kings. Came 1910, and poor Tesla was laughed at and ridiculed, and the people who had supported him financially were forced by mightier players to demand immediate repayment of the invested money. Regarded as a lunatic by the public, deserted by all, and utterly broke, Nikola Tesla died in the United States in 1943.[5]

Clearly, the powers-that-be haven't wanted us to know about these free-energy machines either, but it is inevitable.

The True Motivation of the Greys

Let me return to the Philadelphia Experiment and the real reason the Greys wanted to carry it out. When the Martian atmosphere was on the brink of destruction, the Martians (the ancestors of the present-day Greys) built a complex at Cydonia for the purpose of creating an external Merkaba, a time-space vehicle that would enable them to project themselves into the future. They not only laid out this complex at Cydonia to show mathematically how they did it, but the complex was also the actual mechanism that they used to do it. This occurred at least a million Earth years ago, according to Thoth. The Martians were successful and projected themselves to Atlantis approximately sixty-five thousand Earth years ago.

The descendants of the Martians attempted this experiment again sixteen thousand years ago, but this time they lost control of it, ripping open dimensional levels and pulling spirits from lower levels into Atlantis, where they inhabited the bodies of the Atlanteans.

This failed experiment was attempted in a building that is now on the ocean floor in an area known as the Bermuda Triangle. In this building are counter-rotating star tetrahedral fields creating a huge synthetic Merkaba field.

One of the triangular tips of these fields is sticking out of the ocean, and it is completely out of control.

In order for the Merkaba field to be under control, the counter-clockwise field has to be rotating faster. But the clockwise field occasionally rotates faster, and this is what causes all the distortions. Planes and ships that have disappeared in this area have literally gone into different dimensional levels.

There are also distortions extending into vast regions of space. This is a big problem that extends way beyond Earth.

This problem is also affecting the Greys, so they along with other races are here with permission trying to fix this.

It's not clear how many other comparable experiments the Greys have tried, but they definitely undertook one in 1913. They tried again in 1943 with the Philadelphia Experiment and in 1983 with the Montauk Project. The fields were at least somewhat stabilized in 1993 during a minor peaking of the fields. So even though the fields are now stable, they tried again in 2003 and yet again in 2013.

Al Bielek, who claims to have been involved with the Montauk Project, said the scientists working on the project developed the ability to travel easily in time. They could go as far into the past or future as they wanted. Drunvalo presents the actuality a bit differently. According to him they could go back in time as far as a million years and then they ran into a wall. They could also go forward in time until the year 2013, where they ran into another wall. The reason for this is that the first experiment was one million years ago, so the Greys can't go beyond it, and they also can't go beyond their last experiment, which was in 2013. This is because all of these experiments are interconnected, not just Philadelphia and Montauk, but all of them!

The Greys are trying to solve certain problems and have been given free rein to do so from higher aspects of life. Life is about creating a win-win situation for everyone, not about wiping out the Greys or anyone else; it is about creating something that works for everyone.

Cattle Mutilations

Another aspect of the Greys' experimentation on this planet involves cattle mutilations. They were cutting up cattle to try to understand sexual energy. They stopped being sexual themselves long ago and only continued reproducing via test tubes.

Throughout the 1970s and 1980s, there were thousands of reported cases of mutilations of cattle (and other animals as well) all over the world. Linda Moulton Howe, one of the prime investigators of these incidents, did an ABC television special called "The Strange Harvest" and also wrote a book called *An Alien Harvest*.[6] She first thought that the U.S. government was responsible, but after more research, she concluded it had to be someone more technologically advanced. She found that the incisions were made with incredibly accurate lasers, beyond our present capability. Furthermore, the cows were placed back on the ground with every single red blood cell removed, something we can't accomplish.

The Greys were also abducting humans and conducting experiments on them—in numbers far beyond the limited basis permitted by their agreement with the government. This again was done primarily to understand our sexual and emotional energy. According to Drunvalo, the Greys have now left the Earth and are able to return only for brief periods, so these abductions have mostly ceased.

The Greys now realize that to get out of the trap they created by separating themselves from the source of life, they have to get back their emotional bodies. Unfortunately, the only way they know how to do this is by studying us intellectually, which of course will never work.

Originally the Greys were created much like us, with their emotional bodies intact. It was their allegiance to the Lucifer rebellion that alienated them from the reality of unity. They took external machinery too far, and this just doesn't work. It has been tried many times in the universe, always with disastrous results.

You can create the Merkaba externally. It requires only knowledge, no love, emotion, or feelings. What you get is a UFO. All UFOs, or flying saucers, are based on the principles of the external Merkaba. The problem is this separates you from the source of life, and you eventually lose your emotional body altogether. When you manufacture the Merkaba externally you do not have to use any emotion; you don't have to use any love. Nothing from the emotional body is needed. What this creates is a being who ultimately only knows logic and doesn't have an emotional body.

There is also another problem or fatal flaw with the external Merkaba; there is a limitation on how high you can ascend dimensionally. The great wall of voidness between the octaves was intentionally created to require emotion in order to pass through. No UFOs or any form of external machinery can make it through. This is a detail Lucifer missed—Lucifer, the fourth angel to separate from God, who attempted to base creation on an external template. One million years ago Mars was in the last death throes from the effects of an earlier Lucifer-type rebellion. Many planets and whole areas of space were destroyed as a result. The Martians also succeeded to the point where they had no emotional body or capacity to love. They were based totally on logic and ended up destroying their planet. When they projected themselves in time to Atlantis, they brought their sickness with them to Earth.

The present Lucifer rebellion began in this galaxy about two hundred thousand years ago. The Greys are a quintessential aspect of this rebellion. They now exist on hundreds of planets in the galaxy but are mostly concentrated around Alnilam, the middle star in the belt of Orion. They are related to the Martians, who are among their ancestors. The Greys are a race without an emotional body, and they are dying because of the fatal flaw associated with basing creation on the external Merkaba. They are trapped in their present level of existence. They cannot ascend any higher or go to the next octave of existence because they have separated themselves from the reality to such a degree that they don't even know what love is. This means they can't "feel" their way through creation. They are now aware of this problem; they understand it logically but don't know how to change it.

In their sexual experiments with humans, the Greys attempted to blend themselves or put their essence into our species so that maybe something of themselves would survive. Their initial experiments involving hundreds of thousands of people soon escalated into the millions. These unfortunate ones were taken off the Earth and onto their ships on the fourth dimensional level, where sperm was taken from the males and ova from the females. This was combined with Grey sperm and ova and put into the human females. In three to four months, the fetus was removed and brought onto the ships and raised in test tube-type situations. They cloned another race on this planet through which they hope their genes will continue. As cloning experts, the Greys have done extensive experimentation on humans for a long time. They have made many experimental models of humans in search of the perfect one. They are doing this because they are a dying breed and they know it. Their experiments with humans are an attempt to preserve something of themselves by creating a Grey–Human hybrid. They realize that they are not going to make it, that their particular life form is terminating. The universe has allowed the Greys to do this because all of us on planet Earth are deeply implicated with them from the Martian connection in Atlantis.

There are at least five other extraterrestrial races involved in this drama. For instance, other dimensional beings are making sure the Greys don't overstep themselves as they were wont in the past. In fact, according to Drunvalo, not too long ago the Greys blatantly tried to wipe us out by opening up the fourth dimensional level too quickly. If their attempt had been successful, it would have blown us away.

Accessing higher dimensional levels always needs to happen slowly and organically. When the Greys attempted to force this process, they established a grid around the Earth and were about to open up a dimensional window. The window had the power to work both ways, so in order to make sure the ascended masters of the Earth didn't stop them, they made the window infinitesimally small and opened up the grid randomly at tiny dots all over the planet. However, the ascended masters intuitively interpreted the random pattern, and when the window opened, they projected love back through it and blocked the dimensional hole with light. The Greys were absolutely silent for about three months after this.

This happened in the mid-1980s, and the Greys knew that it was about their last chance here. They saw that our human consciousness was nearing readiness to gain control over what could be called the dreamtime of the planet. This is in fact what happened when the woman from Peru went into Christ-consciousness, raised the ship that was buried under the Sphinx, and cast her spell. The Greys knew when that happened that we would gain control, not them.

A final point regarding the Lucifer rebellion that I intimated earlier but would like to emphasize now is that viewing any of this as good or bad is only our limited interpretation. We are stuck here on the third dimension in polarity consciousness. From a higher level of existence all of this is just an organic process that has both good and bad aspects to it. It has purpose and is leading us somewhere.

14
1972

To begin this story, we must go far back in time to when Thoth and other ascended masters were synthetically creating the Christ-consciousness grid to heal what had happened between the Martians and the Earthlings in Atlantis. When they began their construction, they made the hole in Egypt and began to build the new grid on the old axis. They then activated their geomancy with sacred points all over the world. It was calculated that we would matriculate into the fourth dimension by the winter of 1998, and at that time only a very few people would make it. There was nothing unusual about this experiment of creating the Christ-consciousness grid synthetically. Apparently, it is done on a regular basis—that is, jumping planets up or down dimensional levels. No one in the universe paid any particular attention to this experiment; it was no big deal.

However, about two hundred years ago the Sirians, our father aspect, became aware of a high possibility that we were not going to make it. They foresaw an event that was going to happen in 1972, and they knew that we had to be at the fourth dimensional level by then in order to survive. If we were there by 1972 there would be no problem, but if not, everything would be wiped out—the entire planet. And, as time passed, it increasingly seemed as if we would not reach that consciousness level.

The Sirians did not want to see us eliminated. We are, in a certain sense, their children, and they have that special parental love connection with us. They started searching for a way to solve this problem but were unsuccessful. There was no known way to get a planet at our level of awareness through the changes that were coming. When this had happened in the past it meant that the planet was destroyed. Not to be deterred, the Sirians kept on looking, and eventually they discovered that someone in a faraway galaxy had conceived of an idea that might work but had never been tried. It was still not totally certain that we were not going to make it—consciousness can and does make quantum leaps all the time—but the Sirians were assuming the worst. So, they went ahead and prepared everything necessary to implement the idea.

They created a living vehicle that was fifty miles long, cigar-shaped, black, and seamless with both carbon and silicon life forms on it merged into one. The whole thing was a self-aware living unit. It had a transparent area on one end and was manned by about 300 to 350 men and women of the Sirian race from the third planet. They wore white uniforms with gold emblems. Dedicating as much time to this project as was necessary, they also made eight little flying saucer-type vehicles or ships that were to be unmanned; these were approximately twelve to twenty feet in length. The Sirians worked out all its possibilities, then set it aside and waited.

The actual event happened on August 7, 1972. From our observation—and we didn't know of course that the Sirians had intervened—it was the biggest thing that we had ever seen. Anthony R. Curtis, in his book *The Space Almanac,* called it "the most intense solar storm ever recorded."[1] According to *Science News*: "The early days of August saw a severe disturbance on the sun that produced four major flares between August 2 and August 7. . . . The ones recorded in early August were among the most major ever recorded. . . . The August 7 flare ran the X-ray sensors off the scale. . . ."[2] The solar wind, which has an average velocity of 500 kilometers per second, or 1 million miles per hour, got up to 2.5 million miles an hour for three days, then it dropped down to 1.5 million miles an hour for thirty days. This was considered impossible, yet it happened.

From August to November 1972—a great amount of information was presented about this. Accounts of the event were published in all the major scientific literature of the world and many of the major newspapers, but scientists didn't know what it meant; all they did was publish the data. However, after mid-1973, there was a total worldwide blackout, just as if the great event had never happened.

If the Sirians had not intervened, the explosion would have definitely killed us all. It would have killed everything on the planet right down to the level of microbes and algae. What really happened was that our sun was about to expand in a giant red pulse out to the orbit of Jupiter or thereabouts. For five billion years or so, ours has been a hydrogen sun, where two hydrogen atoms fuse together to make helium. This has been the source of all the light and life on Earth. When the helium that has been building up reaches critical mass, you get another reaction, with three helium atoms coming together to form carbon. At that moment, the sun pulses. A shifting of the sun's poles also took place as a result of a bubble, two-thirds the size of the sun, forming on its surface. If we had been prepared for this, that is, consciously prepared—if we were at the Christ-consciousness level or higher—we could have just tuned to it and it would have been a beautiful trip. But since we had fallen so far in awareness from the events in Atlantis sixteen thousand years ago, we certainly were not prepared.

At the time, 144,000 different races from the other dimensional levels came in here to assist. By mid-January 1972, about 80,000 of them had already arrived. They had a very intense discussion among themselves on the subject of the impending red pulse, and about 79,900 of these cultures said, "There is no hope, there is no possibility of the humans surviving. Let's get out of here." They went home because noninterference was their policy. The other hundred or so cultures—the Pleiadians, the Aldebarans, the Arcturans, and others led by the Sirians—decided to stay and help.

The Sirians not only had the hardware and software in place, they also had ambassadors. The moment the situation was declared hopeless they sent ambassadors to Galactic Command to ask for permission to carry out their experiment. If anyone at all, even just one man and one woman, were to survive this catastrophe, the Sirians would have been refused. But because no one on Earth was going to survive, they did receive permission. They were first asked how many survivors they thought there would be as a result of their experiment. They didn't really know but said, "Probably at least two but no more than ten will make it." A key condition for receiving permission was the belief that at least one person would survive. In truth, as this was a radical thing that had never been tried before, they didn't really know how many would survive.

After receiving permission, the Sirians immediately went to work and within thirty days had everything in place. They launched the large cigar-shaped object just outside the membrane of consciousness of the Earth, at 440,000 miles out, and they placed it one overtone higher so it was invisible to us. They placed the eight small flying saucer-type ships on the apexes of the eight tetrahedral points—that is, the eight points of the star tetrahedron around the Earth. There is a star tetrahedron inscribed in the Earth. There is also a much bigger one, around ten thousand miles above the surface. The points are the chakra system of the planet. Again, these were set one dimensional overtone higher than the Earth. Then from the cigar-shaped ship they shot a beam of laser light, the likes of which we do not have. The beam was about eight inches in diameter, made up of little segments of different colored digital light moving, of course, at the speed of light. This light was coming from one dimensional world into another.

The beam entered into the North Pole and hit the little flying saucer-type ship that was at that tetrahedral point. From there, the immense amount of information was translated into three primary rays—red, blue, and green—that were beamed to the next three ships. These ships repeated this and sent the rays to the next three ships until they ended up at the South Pole. There they were translated back into information and shot into the center of the Earth. From the center of the Earth, by refraction, the information came out in little tiny beams of light by the billions, all over the whole planet. As these beams came out, they connected to all the humans and animals on the planet.

Remember, the Sirians had to protect us from a wall of flame, and this is how they did it. Not only did they have to protect us, they also had to do it in such a way that we didn't know we were being protected. Our knowing would have completely changed the human equation. They set a holographic field around the Earth; then they set up a holographic field around each person and animal. They then began to program events into these fields. In the first few months they didn't change anything, they just got control. They programmed our sky into a hologram and kept everything going as though nothing unusual were happening. They also had to speed up our evolution so we could get to where we could handle being inside the sun. In fact, for a brief time they even took away our free will. We were like little children playing with matches and they said, "No you can't do that." Then they began to program events into our lives so that we would evolve as rapidly as possible. Shortly thereafter, they gave us back our free will, sort of. What they did was to provide us with a set of options and if we picked the wrong one, they would just keep giving us the same set of options until we choose the correct one. At the same time, they were protecting us from the wall of flame.

From the summer of 1972 to the summer of 1974, we were moving in an entirely new direction. We began to accelerate in our evolution (as a personal note, it is exactly in this timeframe, from 1972–1974, that I experienced my inner awakening). Then it started to get out of hand; we began to *really* accelerate. This experiment was much more successful than the Sirians ever imagined. Instead of ten or so making it through, the number is now up to 1.5 billion or so people who will make it through to the next level. We have obviously all been protected from the sun; we're still here. The Sirian intervention also bought time for the synthetic Christ-consciousness grid to be completed. Without this grid, no one can make it to the next level. This grid was completed on February 4, 1989.

Normally, when a planet goes into Christ-consciousness only a very few people initially make it through to live and understand the new reality. The rest drop down to a lower level of consciousness and only over a very long period of time do the initial few pull the rest up until the whole planet reaches Christ-consciousness.

There are, as well, different levels of Christ-consciousness. The levels of Christ-consciousness in the fourth dimension are the tenth, eleventh, and twelfth overtones. It usually takes a long period of time for a planet to evolve through those stages. The first two overtones of the fourth dimension contain the astral plane, where very powerful thought forms have taken on a life of their own. The third overtone is where most people go when they die; the fourth contains faeries and tree spirits. The angelic realms are from the seventh to the ninth overtones. It isn't until you reach the tenth overtone of the fourth dimension that you attain Christ-consciousness. It is to one of these higher overtones of the fourth dimension that we as a planet are headed.

The geometry of the grid around any planet changes as the consciousness of the planet evolves. These changes are normally very rare, usually every thousand years or so. Changes on the grid around the Earth are now happening by the minute. This has attracted the attention of beings from all parts of the galaxy. Because we are inside the system, it is difficult for us to know how fast we are moving. But for anyone outside observing us, it is obvious. What is happening here is unheard of; it has gotten to the point where we are on interdimensional galactic television. That's why the two higher overtones are like a parking lot. Everyone is tuned into us to monitor this event because they know that whatever happens to us here will affect them, too. All life everywhere will be touched by this.

According to Thoth, an analogy for the speed at which we are evolving is one of a newborn baby becoming an adult in fifteen minutes. That was in 1991; now the infant would mature in a matter of a few seconds! This is totally unique; there are no memory patterns for it anywhere, even on the Melchizedek level. It appears to be the most successful experiment that anyone anywhere has ever conducted.

However, the ascended masters don't know what the outcome will be. It thus far appears to be a very successful experiment, but the masters keep figuring out game plans that become obsolete before they can even implement them. Originally Thoth and other ascended masters thought the planet would reach critical mass around the last week of August 1990 or the first week of September 1990, and that by the spring of 1991 we would go into another dimensional level. The masters said they would then bond together, leave the Earth in a ball of light, and go into another level of consciousness. That would pull us up and serve as the trigger for everyone to ascend.

Instead, what happened in August 1990 was that Iraq invaded

Kuwait. The major nations of the Earth did band together, but in preparation for war. Because of this the ascended masters held off. We created planetary unity of a sort by banding together against one man in one country. This was unique in our history. Never before in human history has essentially the entire planet melded together against one person. Even the world wars were very different from this. Because of the Iraqi war, the ascended masters created a new game plan whereby thirty-two of them would go off at one time in a group Merkaba. They actually went through the great wall of voidness and into the next octave. This is how Thoth left on May 4, 1991. That would bring us up a little bit at a time instead of all at once. Every time they do this there will be a rapid expansion in our consciousness. They are timing these events now.

Before Thoth left, he told Drunvalo that he suspects we will not go through the sudden and violent shifting of the poles that usually accompanies a dimensional change. Rather, we will go through a series of steps with our eyes open, and we will do it very harmonically. The ascended masters are going to try to make this a controlled shift of consciousness.

Let's return to late August and early September 1990 for a moment. We *did* reach critical mass. We had the necessary seven to ten percent or roughly five hundred million people who had opened up enough that they would go through. Ascension should have taken place.

The numbers kept growing. We went way beyond critical mass. Then in January 1992 another phenomenon took place—for the first time in sixteen thousand years, the light on the planet was more powerful than the darkness, and it still wasn't igniting. Something else was clearly happening—a cold-fusion process, if you will.

However, because the light was now in control, the Sirians decided we should be given back our power.

The programming of the Earth (how the events take place in our world) was given back to us. Since January 1992 what occurs on the planet is up to us completely. If we could just understand the nature of the situation, that our thoughts and feelings are creating the reality, we could change very rapidly.

15
The Secret Government

Whether you call them the Illuminati, the Cabal, the One Percent, or the Deep State, the name is irrelevant. The "secret government" is basically made up of the richest people in the world. There are a few thousand of them and they have been controlling our outer governments for a long time. They are usually able to control who gets elected, when, and where; they control when there is a war and when there isn't. They control planetary food shortages and whether a country's currency is inflated or deflated. All these things are dominated completely by these people. They can't control natural disasters, of course, but they can and do control a lot.

Somewhere between 1900 and 1930, the Greys made contact with these people. It was obviously well before 1943, because the technology used in the Philadelphia invisibility experiment came from the Greys. Nikola Tesla, a physicist who for a time was the director of the Philadelphia Experiment, stated for the record that he was getting information from ETs, although no one believed him at the time.[1]

In the beginning, this exclusive club thought the Greys were benevolent and made an agreement with them. In fact, they thought they were the best thing that ever came along, a new source of limitless power. The agreement gave the Greys the right to experiment on the planet in exchange for their technology. It is this technology that has enabled our incredible advancements.

This cabal has never shared the Greys' technology with humanity at large, but kept it for themselves. So, there's a two-tiered technology system. There's the tin can type space vehicles that NASA keeps trotting out, and there's the real stuff.

The late researcher Bill Cooper revealed that President Kennedy discovered portions of the truth concerning the alien question. Filmmaker Jay Weidner further asserts that he was informed by many sources inside the military-industrial complex that Kennedy was shown the saucer technology shortly after his election. He went on to say that perhaps the true motivation behind JFK's 1961 speech mandating that NASA land a man on the moon and safely return him before the end of the decade, was to force NASA to reveal the hidden technology, since it was obvious to most everyone in the know that the rocket technology of the day was incapable of getting through the Van Allen radiation belt, let alone getting to the moon and back—that is, with all crew members still alive and functioning.

Weidner went on to say that in order to fulfill Kennedy's ultimatum, NASA hatched a plot to fool the public into believing that we actually did go to the moon—not once, but several times—through the Apollo program. What in fact actually happened, he said, was that filmmaker Stanley Kubrick was brought on board to fake the landings in a movie studio. In return, Kubrick was given a virtually unlimited budget for the filming of *2001: A Space Odyssey* along with the right to make any film he wanted with no oversight. There is a fascinating video called *Room 237: The Apollo 11 Theory* that details the symbolism behind Kubrick's film *The Shining*. After you see that, you might begin to really wonder if the moon landings were real. See: jayweidner. com/interviews

So, the truth be known, they have had UFOs for a long time and are way beyond that. It has been estimated that about fifty percent of the UFOs sighted are our own.[2] These craft are not extraterrestrial but belong to the secret government.

In 1968, scientists in the know predicted that because of strange anomalies on the sun that culminated with the massive explosion on August 7, 1972, that there would be a pole shift sometime around 1984, and that it would demolish just about everything on the planet. The global elite decided to start its preparations to leave. By 1970 they fused the former Soviet Union and the United States. We have been functioning as one country, behind the scenes, for a long time.

They took Soviet, American, and Grey technology and created vehicles to make sure that they could get out of here before the shift. They began to prepare very rapidly, believing they didn't have much time—only about fourteen years according to the latest predictions.

First, they made a base on the moon, using it as a satellite to go deeper into space. They built three small bubble-type cities on the dark side. There was an accident on one of these and many people were killed. Records will indicate that there have been more than two thousand secret missions to the lunar surface.

Once they got enough materials on the moon they went into deeper space, and where do you think they went? Mars, of course, the ancestral home of the Greys. They built an extremely complex underground city designed to hold themselves and a few more people. Not many more, though; their main concern was saving themselves and they didn't much care about anyone else. This has been their style all along. They have become much like the Greys and have lost most of their emotional bodies. They brought to their colony on Mars everything they thought they would ever need.

According to Al Bielek, who worked on the Montauk (officially the Phoenix) Project:

One of the uses of the Phoenix Project, in the use of Time Tunnels, was to provide backup to the Martian Colonies. We went publicly to the moon in 1969. Actually, the Germans were there in 1947. And we went there in 1962 with a joint U.S.-Russian expedition. Then

we went to Mars on May 22, 1962. The movie, *Alternative 3,* done
by Anglia Television, April 1, 1977, which is available in the underground, outlines it completely. It shows the actual transmission—
the color shots by TV back from Mars, as this *Explorer* moved and
landed.[3]

As a quick sidenote, I'm sure you noticed what he said about the
Germans going to the moon in 1947. I simply did not believe this to be
true; after all, they lost the war—right? As it turns out, that statement
appears to be absolutely accurate. Yes, they did lose the ground war, but
they won the space war. I will leave that as a mystery for the moment;
however, I wrote about this in 2023's *Catching the Ascension Wave.* I
will also say that Al Bielek is one guy who has stood the test of time. He
has proven to be a reliable source of some rather amazing information.

At that time not only did this cabal not want anyone to know what
they were doing, but they didn't want anyone to compete with them
either. Anyone who was involved in advanced technology was stopped
in one way or another (as with a gag order). If they couldn't stop them,
they just got rid of them.

Somewhere around 1984 the city on Mars was completed. The
global elitists began thinking that they really had it made. Then just
a few years later, approximately 1989, they made a shocking discovery. The Earth is not the only place that the north and south poles are
shifting; it is happening on all the planets in our solar system including
Mars!

Seven to nine months later a further blow hit when they learned
that it wasn't just a physical change that was occurring but also an interdimensional shift in consciousness. At that point, like the Greys, they
felt quite helpless. They then began to realize that, like it or not, they
are connected to those "useless eaters" down there—all of humanity.

This was the situation the Earth was faced with in Atlantis that
was ignored sixteen thousand years ago when the Martians tried to split

off from the rest of the Earth's population and go their own way. The global elite are now at least beginning to understand that they can't survive on their own.

Insofar as these people are totally aligned with the Greys, they are pretty much devoid of emotions; however, they are extremely intelligent. If they could figure out any other way to save themselves, they probably would, and that appears to be the way they are going. I will have more to say about their possible fate in the next chapter. But on some level, they are now realizing that if they are to survive, we all must—we are all in this thing together.

This is what the higher beings have wanted since the disaster in Atlantis. They didn't want just the Earthlings to survive at the expense of the Martians or vice versa; they wanted them both to survive and to go from there.

16
End-Time Prophecies

The American Indians say that we're going from the Fourth to the Fifth World rather than from the third to the fourth dimension. That's because they see the Great Void as a world.

The following from *Book of the Hopi* gives a pretty good description of the Void:

> The first world was Tokpela (Endless Space).
>
> But first, they say, there was only the Creator, Taiowa. All else was endless space. There was no beginning and no end, no time, no shape, no life.
>
> Just an immeasurable void that had its beginning and end, time, shape, and life in the mind of Taiowa the Creator.[1]

Emergence into the Fourth World

The Hopi believe that their people came from within the Earth. They say it has happened three times before when the world became polluted and immoral and they had to leave. Each time they went up into the sky, which they say hardens at one level and they come out onto a new surface of the Earth.

The Hopi, according to the late Dan Katchongva, son of the great leader Yukioma, dwelled in a Third World inside the Earth for many years. "But eventually evil proved to be stronger. Some people forgot or ignored the Great Spirit's laws and . . . began to do things that went against his instructions . . . This resulted in a great division, for some wanted to follow the original instructions and live simply."[2]

It finally reached the point where people had no respect for anything. Life had become *koyanisqatsi*—a world out of balance.[3]

According to Hopi emergence myth, the leaders then gathered to pray and ask for guidance. The idea came to them to move. They had been hearing footsteps in the sky, and then "a *tootsa,* or small sparrow hawk, found the *sipapuni,* or hole in the sky, and emerged into the Fourth World."[4]

The bird found Maasaw, Caretaker of the Earth, who eventually granted permission to enter the Fourth World on the condition that the people practice his way of life, and the evil ones stay behind.

> The sparrow hawk relayed the message and those waiting were overjoyed. But being unable to fly, how would they get to the upper world? Again they smoked and prayed for guidance, then called upon the spirits of nature for help. Inspired, the *kikmongwi* and the priests planted a spruce tree. A squirrel brought them pine seeds and they were planted. Then the leaders prayed and sang the creation songs. The trees shot into the air. But the branches were too soft to support the people all the way into the upper world. So, a bamboo reed was planted.
>
> All this was kept secret from the corrupt majority of the population. Only good-hearted people were informed of the plans to leave. . . . They began to climb up on the inside of the reed, resting between the joints as they worked their way up. The bamboo did not have sections in it at the beginning. The sections were created as the people rested during their long climb skyward. Finally the pointed end of the bamboo pierced the sky and the people climbed through the emergence hole, or *sipapuni,* into this, the Fourth World.[5]

This is the fourth time the world has become polluted and immoral, and the Hopi are ready to leave again—what they call the "Day of Purification"—and they will go into the Fifth World.

White Feather

The following Hopi prophecy was first published in a mimeographed manuscript that circulated among several Methodist and Presbyterian churches
in 1959. The account begins by describing how while driving along a desert highway one hot day in the summer of 1958, a minister named David
Young stopped to offer a ride to an Indian elder, who accepted with a nod.
After riding in silence for several minutes, the Indian said:

> "I am White Feather, a Hopi of the ancient Bear Clan. In my long life
> I have traveled through this land, seeking out my brothers, and learning
> from them many things full of wisdom. I have followed the sacred paths
> of my people, who inhabit the forests and many lakes in the east, the
> land of ice and long nights in the north, and the places of holy altars of
> stone built many years ago by my brothers' fathers in the south. From
> all these I have heard the stories of the past, and the prophecies of the
> future. Today, many of the prophecies have turned to stories, and few
> are left—the past grows longer, and the future grows shorter.
>
> And now White Feather is dying. His sons have all joined his
> ancestors, and soon he too shall be with them. But there is no one
> left, no one to recite and pass on the ancient wisdom. My people
> have tired of the old ways—the great ceremonies that tell of our ori-
> gins, of our emergence into the Fourth World, are almost all aban-
> doned, forgotten, yet even this has been foretold.
>
> The time grows short.
>
> My people await Pahana, the lost White Brother [from the stars],
> as do all our brothers in the land. He will not be like the white men
> we know now, who are cruel and greedy. We were told of their com-
> ing long ago. But still we await Pahana.
>
> He will bring with him the symbols, and the missing piece of
> that sacred tablet now kept by the elders, given to him when he left,
> that shall identify him as our True White Brother.

The Fourth World shall end soon, and the Fifth World will begin. This the elders everywhere know. The Signs over many years have been fulfilled, and so few are left.

This is the First Sign: We are told of the coming of the white-skinned men, like Pahana, but not living like Pahana-men who took the land that was not theirs. And men who struck their enemies with thunder.

This is the Second Sign: Our lands will see the coming of spinning wheels filled with voices. In his youth, my father saw this prophecy come true with his eyes—the white men bringing their families in wagons across the prairies.

This is the Third Sign: A strange beast like a buffalo but with great long horns will overrun the land in large numbers. These White Feather saw with his eyes-the coming of the white men's cattle.

This is the Fourth Sign: The land will be crossed by snakes of iron.

This is the Fifth Sign: The land shall be crisscrossed by a giant spider's web.

This is the Sixth Sign: The land shall be crisscrossed with rivers of stone that make pictures in the sun.

This is the Seventh Sign: You will hear of the sea turning black, and many living things dying because of it.

This is the Eighth Sign: You will see many youths, who wear their hair long like my people, come and join the tribal nations, to learn their ways and wisdom.

And this is the Ninth and Last Sign: You will hear of a dwelling-place in the heavens, above the earth, that shall fall with a great crash. It will appear as a blue star. Very soon after this, the ceremonies of my people will cease.

These are the Signs that great destruction is coming. The world shall rock to and fro. The white man will battle against other people in other lands—with those who possessed the first light of wisdom. There will be many columns of smoke and fire such as White Feather has seen the white man make in the deserts not far from here. Only those which come will cause disease and a great dying. Many of my people, understanding the prophecies, shall be safe. Those who stay and live in the places of my people also shall be safe. Then there will be much to rebuild. And soon—very soon afterward—Pahana will return. He shall bring with him the dawn of the Fifth World. He shall plant the seeds of his wisdom in their hearts. Even now the seeds are being planted. These shall smooth the way to the Emergence into the Fifth World.

But White Feather shall not see it. I am old and dying. You perhaps will see it. In time, in time. . . ."

The old Indian fell silent. They had arrived at his destination, and Reverend David Young stopped to let him out of the car. They never met again. Reverend Young died in 1976, so he did not live to see the further fulfillment of this remarkable prophecy.

The signs are interpreted as follows: The First Sign is of guns. The Second Sign is of the pioneer's covered wagons. The Third Sign is of longhorn cattle. The Fourth Sign describes the railroad tracks. The Fifth Sign is a clear image of our electric power and telephone lines. The Sixth Sign describes concrete highways and their mirage-producing effects. The Seventh Sign foretells of oil spills in the ocean. The Eighth Sign clearly indicates the "Hippie Movement" of the 1960s. The Ninth Sign was the U.S. Space Station Skylab, which fell to Earth in 1979. According to Australian eyewitnesses it appeared to be burning blue.

The Blue Star Kachina

Frank Waters in *Book of the Hopi* explains how an individual who obeys the laws and conforms to the patterns of perfection given by the Creator becomes a *kachina* when he dies. This enables him to go immediately to the next universe (there are seven in total), thus avoiding the intermediate worlds. He then returns periodically to help humankind continue its evolutionary journey.

The kachinas are inner forms, the invisible spiritual forces of life who act as messengers. Their chief function is to bring rain, ensuring the abundance of crops and the continuation of life.

According to Waters:

Kachinas are properly not deities. As their name denotes (*ka,* respect, and *china,* spirit), they are respected spirits: spirits of the dead; spirits of mineral, plant, bird, animal, and human entities, of clouds, other planets, stars that have not yet appeared in our sky; spirits of all the invisible forces of life.[6]

Kachinas are impersonated by men in dances wearing masks, who lose their personal identities as they are imbued with the spirits they represent.

According to Hopi prophecy, when the Blue Star Kachina makes its appearance in the heavens, the Fifth World will soon emerge. The late Dr. Robert Ghost Wolf was told the story of the Blue Star Kachina by grandfathers when he was very young. According to Ghost Wolf:

It was told to me that first the Blue Kachina would start to be seen at the dances, and would make his appearance known to the children in the plaza during the night dance. This event would tell us that the end times are very near. Then the Blue Star Kachina would physically appear in our heavens which would mean that we were in the end times.[7]

He tells how in the final days the Blue Star Kachina will join with Poganghoya and Palongawhoya, the twin guardians of the north and south poles, respectively, and that they will return the Earth to its natural rotation, which is counterclockwise.

According to Thoth, the Earth's rotation is indeed reversed every time we go through a dimensional shift. Having experienced this four times previously, he states that he has seen the sun rise first in the West, then in the East, the West, and the East (where we are now). It will return to the West after the next shift.

Ghost Wolf continues:

The return of the Blue Star Kachina who is also known as Nan ga sohu will be the alarm clock that tells us of the new day and new way of life, a new world that is coming. This is where the changes will begin. They will start as fires that burn within us, and we will burn up with desires and conflict if we do not remember the original teachings and return to the peaceful way of life.

Not far behind the twins will come the Purifier, the Red Kachina, who will bring the Day of Purification. On this day the Earth, her creatures, and all life as we know it will change forever. There will be messengers that will precede this coming of the Purifier. They will leave messages to those on Earth who remember the old ways . . .

Those who return to the ways given to us in the original teachings and live a natural way of life will not be touched by the coming of the Purifier. They will survive and build the new world. Only in the ancient teachings will the ability to understand the messages be found . . .

Nothing living will go untouched, here or in the heavens. The way through this time it is said is to be found in our hearts, and reuniting with our spiritual self. Getting simple and returning to living with and upon the Earth and in harmony with her creatures. Remembering that we are the caretakers, the fire keepers of the Spirit. Our relatives from the Stars are coming home to see how well we have fared in our journey.[8]

Hopi Prophecy Fulfilled

Comet 17P/Holmes shocked astronomers on October 24, 2007, with a spectacular eruption. In less than twenty-four hours, the seventeenth-magnitude comet brightened by a factor of nearly a million, becoming a naked-eye object in the evening sky. By mid-November the expanding comet was the largest object in the solar system—bigger even than the Sun. Since then, the comet has faded back to invisibility.[9]

This exploding "Blue Star" was the fulfillment of a two-hundred-year-old Hopi prophecy, opening a seven-year window, putting us in the "end times." The Mayans are in complete agreement with the Hopi, as both are conscious survivors of Atlantis. They both hold the memory of the Earth for the past twenty-six thousand years. The Hopi, in fact, used to be Mayan. This was told to Drunvalo by Grandfather Eric, the last living member of the Bluebird Clan (the historians of the Hopi Tribe). He said that they did not come across the Bering Strait; they came from Guatemala, and a decision was made long ago that a group of Mayans would head north to find a new place. That group became the Hopi.[10]

This means that if the prophecy was to be taken literally, that sometime between late 2007 and 2015 or so, we were very likely to experience changes almost greater than our ability to imagine: a physical pole shift, along with a consciousness shift into the fourth dimension. Now since that didn't happen in the predicted window, does that mean that it will not happen, or was the timing simply off?

In case you're wondering about the Red Star Kachina—the one that according to the prophecies, will usher in the Day of Purification—Betelgeuse is a likely candidate, since it's already near the end of its lifespan and is expected to explode as a supernova. However, this could be anywhere from now to many years into the future. When this massive star, nine hundred times the size of our sun does explode—and even though it is six hundred and forty-two light years away—it will appear as bright as the moon in the night sky for several weeks and may even be visible during the day.

The Mayans remember the last shift, along with the one twenty-six thousand years ago. They say many millions of people died when Atlantis went down just because they didn't know a few simple things. There are certain internal changes we must make in order not only to survive, but also to rise to the next level. The keys, they say, are to stay out of fear, to be calm and balanced, and to be in the heart. They say Mother Earth knows the vibration of the heart and that she will protect you through these times.

The Mayans also say that of everything written about them in the past five hundred years, none of it came from them. It is their intention to rewrite their history and their knowledge, and everything that is known about them over the past twenty-six thousand years. They are about to put this into a book.

The Event

The latest prophecy that is making its rounds on the internet—though similar in many respects to the information I've been hearing for the past thirty years, beginning with The Harmonic Convergence, in 1987, then the alleged arrival of the photon belt in the early 1990s, and followed by NESARA—is known as *The Event*.

So, what is The Event? Well simply put it is an updated version of an old story telling of a singular cosmic "explosion" of Source energy coming through the Great Central Sun. It will according to a channeled source known as Sananda (channeled through Adele Arini), arrive and encompass the entire Earth in a single moment—reaching every living being on the planet at the same time. It will do so, he says, in a way that not one single person will remain unaffected. He goes on to say that all incarnate souls will feel powerful, undeniable surges of unconditional love, divine bliss, complete acceptance of who you are, divine grace/blessings, and strong feelings of finally being home again.[11]

Then what will happen after The Event—since everyone will perceive and experience it differently—depends upon which of the four soul groups you fit into.

Group one consists of those who are ready and have been waiting and preparing for ascension for some time. This "first wave" will be the guides, the way-showers and the leaders of fourth-dimensional living. According to Sananda, many new, fourth-dimensional structures are now ready to be put into place and for those in this group, your spiritual powers will be fully awakened and start to develop at a faster rate. You will learn how to master your thoughts since humanity will soon be a race of beings who, predominantly, communicates via telepathy. Mastering your thoughts is the first step to mastering the art of creation and manifestation. And your galactic brothers and sisters will greatly help you on this as well. And when your lessons with them are complete, you will then go out into the world to share this knowledge with others so that all humans on planet Earth will one day be the living embodiment of their Higher Selves. All will become powerful creators and manifestors. Earth will become a planet that only consists of loving beings who truly embody Oneness and Unity; of masterful beings who have ascended to the fourth dimension.

Group two consists of unawakened souls (who, at a higher level, have decided to ascend in this lifetime). For this group, The Event will trigger tremendous shock, confusion, and cause them to initiate research into greater understanding on what exactly has happened to them. The Event will be the loudest wakeup call they have ever heard in their entire lives; one that can no longer be ignored. They will seek and attract into their lives, guidance from already awakened light workers; spiritual development will become their main focus in life. And with time as they grow more spiritually mature, these souls will also come into a full remembrance of their true nature, and strive to embody divine love in their everyday lives.

Group three consists of unawakened souls (who, at a higher level, have decided not to ascend in this lifetime). They will still feel the intensity of The Event and also experience all the feelings that come during it. However, after it is over, they will slowly return to life within the third-dimensional paradigm. Even if they were to come across real and authentic news about The Event and what it signifies, their inner reaction will be strong doubts, disbelief, and denial.

Group four consists of unawakened souls (who, at a higher level, have decided to play the roles of the dark forces in this lifetime). They will come to a sudden realization that their time has now come to a complete end. They will know that all is lost for their dark cause, as humanity will no longer hold any interest in living in the type of lower-vibrational, third-dimensional consciousness these dark forces thrive in.

All the souls in group four will be faced with only two possible paths to undertake after the Great Event.

Path A is the path of a possible return to the light and love of their Higher Selves. For those who are choosing this, great courage will be needed to walk the path. They will all have to "come clean" with all the wrongs they had done in their current lifetime (i.e., all the things that were harmful to the planet, and/or harmful to the people.)

Path B is for those who refuse to repent. They will have to be "transitioned" back to the Light of Source—back to God's loving embrace in the spirit realm—to then be reincarnated to a different third-dimensional planet that still supports duality, fear, lack, and separation consciousness. Their return to the Light of Source will happen in a loving, respectful manner. These souls will be allowed to continue to engage their own path by living in another low-vibrational planet and play the roles of the dark forces until they are ready to reunite with the light and love of their Higher Selves.

So, when will this Great Event happen? That's hard to say; it could be as soon as now, or somewhere off in the not-to-distant future. With that being said, the most important thing to bear in mind is that it, or something similar to it, will occur. Mother Earth as I will detail in later chapters, has already made all the necessary preparations. In the grand scheme of things, anytime a third-dimensional planet has gone this far, it has always made it to the next level, which is the tenth, eleventh, or twelfth overtone of the fourth dimension. There is nothing left to prevent it!

17
The Shift of the Ages

Okay, I have made the case throughout this book that we are heading into changes almost greater than our ability to imagine. We are about to enter into a new way of being that is—shall we say, one hundred or more times more harmonious than even our very best moments here on third-dimensional Earth. Take a moment now to imagine, what would your life be like if you were one hundred times more present, joyful, and grateful. One hundred times more inspired and creative, and so on. Would that be worth shooting for?

At the moment, we are all in transition, we are all going through tremendous change. For many of us, our old world is falling apart while the new one is not yet firmly in place. It also means that we are in the process of completing the past. We must clean up the mess we have made before we can make this move. My sole purpose is to make it my personal business to see to it that our journey from one dimension to the next is a safe, smooth ride.

Many people interpret these changes metaphorically. For example, on a radio program where I was recently interviewed, one caller had his own version of California falling into the ocean, saying that it had already happened: in 1964 Esalen Institute opened and California did fall into the ocean—the ocean of emotions—and it started a revolution toward the light. He also had his own interpretation of Edgar Cayce's prediction of the destruction of New York. "What is New York?" he asked. "It is the financial center of the world. This means the end of the material world will come to pass. We are going towards the light with this whole evolution of consciousness. People have to think of these things as metaphors."

Another individual (whom I'll call David) sees the upcoming pole shift as a polarity shift, a balancing of the masculine and the feminine principles—much overdue balancing, by the way. He understands it as a metaphor for profound internal changes. The masculine or the left-brain in all of us, the part that sees itself as separate from and trying to control Nature and life, is being balanced by the feminine or right-brain, the side in all of us that is able to intuit the oneness of all life.

David believes the Earth is already in the fourth dimension and that humanity is fluctuating between the third and fourth dimensions. He sees the third dimension as a high-level dense place where fear, duality, and separation appear to be real. As we move up in vibration, these elements begin to dissolve—first on an inner level and then on an outer one. This is extraordinary, as he sees it, because the whole socio-economic system—the medical military industrial complex, as he calls it—is a fear-based system. It is based on keeping people disempowered—that's how they control us. Since we have bought into it, we have simultaneously accepted a very limited conception of who we are. That is all changing. As it changes, we are stepping into our birthright of health, peace, and freedom. We begin to identify much more closely with our true nature.

I am in complete agreement with all of these points—and I see the upcoming shifts literally, that we really will switch dimensional levels into a shorter wavelength where the reality is completely different, and that the poles will shift at the same time. I also see that these events are directly related to our consciousness. If we approach these changes in fear, then we will create a fearful experience, a cataclysm. If we can raise our awareness to the point where we can attune to these events, they will be a beautiful experience, a different kind of cataclysm. Since it is not helpful in any way to approach the coming shift with fear, we must find a way to integrate any fears we may have about the coming changes. Stay tuned—that is the main theme in the remainder of this book.

But for now, we need to dig deeper and see if we can get a better grip on the rather extraordinary times we are in. Let's begin by taking a look at Venus.

Venus

Venus today—on the third dimensional level—is a rather inhospitable place, and that's putting it mildly. A quick online search reveals an average temperature on Venus of 864 degrees Fahrenheit, or 462 degrees Celsius. There's sulfuric acid in the atmosphere, which is from volcanic eruptions. There's a very, very thick atmosphere; you can't even see the surface of Venus from space.

However, on the fourth dimension, Venus is a well-populated, beautiful planet. It is peopled by the Hathor race, the most intelligent and advanced race by far in our solar system. They are far more advanced than humans, the Greys, or the Nefilim.

The Hathors are Christ-consciousness beings. They stand ten to sixteen feet tall and base all their science on sound currents coming from the throat. They are beings of pure light and tremendous love. They have been working with the Egyptians and with us for a long time.

It was approximately 2.6 billion years ago that Venus graduated and became fourth dimensional. At that time, its total population, according to Ra and *The Law of One*, was only thirty-eight million. Evidently, conditions on the third dimensional level were a bit harsher than what we experience on Earth.

And the Earth is no longer going to be habitable for people on the third dimension for very much longer. If we want to stay with the Earth, we're going to have to upgrade to the higher overtones of the fourth dimension and transform into light bodies to be comfortable, and at that level it's going to be fine—in fact, as I said earlier, our lives will be a hundred times more harmonious than even our best moments here. We will soon realize that our new home is truly a heaven on earth!

The 6.5 million Venusians who made it to the next level became known as the Ancient Builder race. Not only did they establish a colony in Antarctica, but they also ventured throughout our local sector of space and built crystalline towers, obelisks, domes, and pyramids. These structures—built with a glass-like structure called transparent aluminum—created vast habitable regions inside of various planets and moons, including our own. This information came to us from whistle-blowers inside the Secret Space Program.

Of the total population on Venus of thirty-eight million, about six and one-half million people made it into the higher level. The remaining 31.5 million inhabitants had to repeat life on the third dimension elsewhere. So, the question is, where did they go, and how did they get there?

There is a most interesting book called *Cosmic Voyage*, where the author—Dr. Courtney Brown—through remote viewing, saw how it is done. If you question the validity of remote viewing, you should know that with a well-trained person, this technique is extremely accurate—about ninety-nine percent.

It is apparently a common occurrence that only a portion of a planet's inhabitants make it into the fourth dimension. Those remaining are not left out to twist slowly in the wind, or more accurately, to directly experience the shifting of the poles. They are transported to another third-dimensional planet to give them the opportunity to alleviate enough of their karmic debt, so they will eventually become ascension ready. I gave the full details on the fascinating aspects of this transfer in *Catching the Ascension Wave*.

What the Shift Might be Like for Us

Because our situation is so unusual no one—including the ascended masters, and Drunvalo—can say for sure what our experience will be like. Thoth does believe, however, that we will likely have a much more gentle, organic, and conscious ride from one dimension to the next than is usually the case.

Drunvalo says he does know what usually happens when the poles shift. When we approach the point in the precession of the equinoxes where the change takes place, everything begins to break down.

The key is the magnetic field of the Earth, which is what we use to interpret who and what we think we are and also to store our memory. We are very much like a computer in this sense. We need some form of magnetic field to process data.

The day before, the day of, and the day after the full moon there is usually an increase in murder, rape, and other crimes because the moon causes a bubble on the magnetic field. This minute bubble is enough to push already emotionally disturbed people over the edge.

As the shift approaches, things start to get out of balance, and the magnetic field begins to fluctuate significantly over a very short period of time (about three to six months). This would be like a full moon getting bigger and brighter every day. What happens then is that people start to lose it emotionally. This breaks down economic and social structures on the planet because it is only people who are keeping it all together. When they lose control, everything else falls apart.

This may not happen this time. We may be able to keep the magnetic field together.

Usually, the pole and dimensional shifts are simultaneous. About five to six hours before the actual shift happens, an extraordinary visual phenomenon takes place. The third and fourth dimensions actually begin to interface. Third-dimensional consciousness gradually recedes from us as we approach the fourth dimension.

As the third-dimensional grid begins to break down, synthetic objects disappear. This is one reason that even though there is a five-hundred-million-year history of advanced life on this planet, there is virtually no evidence of it. In order to survive pole shifts, objects must be made purely out of natural materials like the pyramids and the Sphinx—natural materials that are in resonance with the Earth. Even then virtually everything on the planet is literally blown away.

As these objects begin to disappear, fourth-dimensional objects may suddenly emerge. Colors and shapes unlike anything we have ever known will appear on the landscape.

Even if we are able to keep the magnetic field together up until the actual shift, there will almost certainly be a brief period when it will be completely gone. As soon as this field collapses, the Earth will disappear for you, and you will be in the Great Void. The duration of this "bardo" is three and a half days. Then life will come back in the fourth-dimensional world.

You will find yourself in a brand-new world the likes of which you have never conceived. You have been there many times before, but your memory of it has been erased. You will be just like a baby, having no idea what all the colors and objects are.

One similarity between this world and the next is the Holy Trinity—that is, the mother, father, and child. As you enter this new world, though you understand nothing, you will be greeted by two beings, one male and one female, father and mother. You will have a very close love connection to them. When you enter the fourth dimension, it will take about two years experientially to grow and mature. The growth is literal, by the way, given that the height range on this level is from ten to sixteen feet. Your new parents will guide and assist you during your growth.

You will appear just as you are now, though more than likely naked (clothing usually doesn't make it through). However, the atomic structure of your body, though it comes through, will have changed dramatically. The mass of your atoms will have been converted to energy. The individual atoms will have separated from one another at phenomenal distances. Most of your body will be energy, a light body. Remember, in this new world, you will be creating your reality moment by moment with your thoughts. Manifestation will be instant: if you think "orange," an orange will instantly appear and you can peel and eat it.

This is why thoughts like peace, beauty, love, and so on are so important. If you are motivated by fear in the fourth dimension, you will create and manifest your reality instantly, and you will find yourself confronted by something terrifying like an antagonist trying to murder you. Then you will instantly manifest a gun and shoot them, and when that happens you will be bounced right back down to the third dimension where the link between cause and effect is slower. The quality of thoughts is totally important on the fourth dimension. This is what Jesus meant when he stressed the purity of thoughts. Love and peace and unity and being kind to your neighbors are ultimately very practical because they work reciprocally.

These things are important on the third dimension also, but here it's easier to get away with playing dumb and not seeing cause and effect. The third dimension seems to be a realm for mastering limitations or victim consciousness. In victim consciousness, the ultimate victim is one who doesn't know that he or she is creating reality, and believes that things just happen. That however, is now changing as we've turned the corner and are now heading back into the light. The game plan is changing from mastery of limitation to mastery of divine expression.

18
Our "Illusory" Holographic Universe

In our current state we tend to think that we live inside this body, that everything "out there" is separate from us, and that our thoughts and feelings don't manifest beyond ourselves. We think that we can hide our feelings and thoughts and they have no impact "out there." This is simply not true. Everything that we think, feel and do is creating our entire reality, clear out to the most distant stars. We are creating everything in every moment, more so than we could ever imagine.

The statement about creating our reality clear out to the most distant stars is literally true by the way. Stay with me on this and I will show you exactly how you and I are creating our reality in this "illusory" holographic universe.

By the way, much of the following information came from the video of a lecture called *The Lion Sleeps No More*. This eight-hour presentation was given at London's Brixton Academy by David Icke. David's purpose is as quoted from the back cover of the DVD set: He takes the manipulation of the human race and the nature of reality to still new levels of understanding and he calls for humanity to rise from its knees and take back the world from the sinister network of families . . . that covertly control us from cradle to grave. In order to take back our power, it sure helps to know who we really are; in this seminar, David did a masterful job of showing us our true nature as Infinite Consciousness having a human experience.

To begin, take a look outside your window—what do you see out there? A tree perhaps, and maybe a couple of squirrels running around having a good old time. But is what you're looking at really "out there?" My research says not; it tells me that this reality, in terms of its physicality, is an illusion. There is no physicality; there is no "out there." Everything you are looking at "out there" is going on inside of you. In fact, it is merely a holographic projection of your consciousness.

Albert Einstein clearly understood this when he said "*Reality is an illusion, albeit it, a persistent one.*" The reason it's persistent is because we live in a virtual reality universe, an incredibly expanded, advanced version of a gigantic computer game.

It's becoming much easier to talk about reality, because technology is starting to mirror the very reality that we are experiencing. It's getting closer and closer to real, and the projection is that in the near future, there will be computer games in which you can hardly tell the difference. There are training simulations used in corporate situations to improve business awareness and management skills. Driving simulators are being increasingly used for training drivers all over the world and flying simulations are used to train pilots. Some hospitals are using virtual reality images to show very cold images to burn patients. The brain decodes that and cools the skin down.

The advanced virtual reality technologies are all hijacking the way the five senses work. They are feeding digital information to the brain, and tricking it into believing that something is going on that is not.

We have the biological body computer, it has the ability to not only react to data, but also to asses that data and make decisions on it.

The base foundation of this virtual reality universe is waveform, which is vibrating energy. Within waveform, incredible amounts of information can be stored; and that's what the base information construct of this virtual reality is—it's information in waveform. What happens is the waveform information construct is decoded through the body computer into the world that we think we are experiencing; but it's all going on in your head.

It's almost a mirror, although much more advanced, of the wireless internet, where through a computer, you can pull the World Wide Web out of the unseen to appear on a screen anywhere in the world.

What we're doing then is decoding vibrational information into electrical information, which is sent to the brain. And the brain decodes that into the world that we think we are experiencing. The five senses change vibrational information into electrical information which goes to the brain to be decoded into holographic information. So, the five senses are a decoding system; the most obvious one is sound—a vibration comes to the ear and sends a message to the brain, and the brain hears the sound. There is no "sound" until we have decoded it as such—sound is just a vibration until then. It's the same with taste, electrical signals are sent to the brain that are decoded. An apple tastes like an apple only after it has been decoded as such. It's the same with the other senses. With sight for example, you will only see a tree after it has been decoded. You may have heard the story of how the indigenous people of the day had no frame of reference for what was happening, and thus were unable to "see" the three ships of Christopher Columbus. This decoding phenomena is even true with movement, there is no movement until we've decoded that movement. It's all an illusion!

So, we live in a very advanced equivalent of the World Wide Web. The only place the internet exists in a usable form is on your computer screen. Everywhere else, it's electrical circuits. The only place television exists as moving pictures is on your television screen. Everywhere else, it's electrical circuits and broadcast frequencies. And only after we have decoded it, does our apparent "out there" reality appear on our "screen." Any other time, it's a vast information construct.

We live in the vastness of the cosmos, where information in the energy around us is pulled out of the unseen. The same is happening on your computer with Wi-Fi (no wires, no connections), you are pulling the unseen onto your computer screen.

If we all are creating the reality, then why do we all see the same things, the same car and the same tree and so on? When you log onto the internet, no matter what part of the world you're in, you log onto the same collective reality as everyone else. You can go to the same website no matter where you are. You can put your individual spin on it—whether you like it or not, just as you might like a movie you've seen and someone else might not like it—but it's the same collective reality.

And the body computer is pulling this collective reality out of the unseen, just like your computer. Because we are all decoding the same basic construct, that's why we see the same things.

We are picking up information from the cosmos and decoding it into our experienced reality. What we do and do not decode from it, is dictated by how our decoding system works, and that is dictated by our sense of perception. And our belief systems come from our sense of perception.

For the past thirteen thousand years, we have come into this world in a state of separation, cut off from the source. Our beliefs then, are based in fear and limitation, and survival, coming from our limited sense of perception.

What you believe about life is what controls your life. Most of these beliefs are on a subconscious level. The brain decodes reality through these beliefs, and it becomes a self-fulfilling reality. This virtual reality universe then becomes a virtual lie. We are *not* our identity, we are infinite consciousness having a human experience.

What this universe is then, is information; and it is information decoding information. Just as you put a computer disc into a computer—its information. The information encoded in the computer then reads the information on the disc. That's what we are doing, and that's how we manifest this apparently "out there "reality. Everything "out there "certainly seems to be outside of us, but it's inside of us, we are creating it through the decoding system in our head, heart, and genetic structure.

Science tells us that there is a massive black hole at the center of our galaxy. It is increasingly believed that there is a giant black hole at the center of every galaxy. These black holes resonate the base vibrational frequency of this virtual reality universe, and they interact with the suns. It triggers information being transmitted by the suns in the form of photons. Photons are the base unit of light and all other forms of electromagnetic radiation; they carry information. These photons, being generated out of the suns, triggered by this vibrational base frequency coming from the black holes, create the information in the energy around us within our reality.

That's what we are decoding, the Earth decodes it, and what it's decoding is photon information that is passing through the Earth's energy grid. And prana is photon information that the body computer is decoding into this collective reality.

Manifesting from All Possibility

Miracles are simply overcoming the programming of what's possible and not possible. A miracle to one person might seem like ordinary stuff to another. We for example, are programmed to believe that you will get burned if you walk through fire. If you walk through the fire with that belief, that decoding state, you will burn your feet. But if you can go into another state of consciousness and override that, you can walk across hot coals and not feel a thing.

There are no miracles; there is just understanding how to manifest all possibility. That understanding comes from acknowledging that we are consciousness, disembodied, no form awareness, having a human experience.

The body is the way our awareness experiences this reality. If you wish to interact with this reality, which is a frequency range, you have to have an outer shell that vibrates within this range. Consciousness (your true self) is vibrating much too quickly, it's on a different wavelength. So you take on this outer shell, which we call a body. That enables you to interact with this reality.

If you believe that vehicle (body) is who you are, then you go from all consciousness to your individual identity. You become a powerless victim. What this does is puts us in a false identity, one that doesn't think all possibility—it thinks limitation. You are *all* consciousness, *all* awareness, *and all* possibility. You are a Spiritual Being having a human experience.

All possibility, beyond vibration (God on the other side of the wave form universe) is stillness and silence. If it vibrates, it's illusion; it's the wave form universe. All possibility is everything and nothing. Out of nothing comes the realm of form.

There was a time before any of these wave form universes existed. All there was, was the Eternal Oneness in perfect awareness of itself. Then God decided to create this multidimensional reality in order to give Oneness an opportunity to experience all the possibilities offered in waveform. He/She split him/herself into two. Part of Him/Her stayed on the other side, while His/Her other half stepped into the newly created worlds.

Out of all possibility, there is virtually an infinite number of ways to experience this wave form universe, all the different dimensional levels. And as I suggested in chapter 5, they all share the same space.

We live in a frequency range, and that range is dictated by the range of frequencies that our body—computers can decode. And it is extraordinarily tiny, we are virtually blind in terms of all that exists. The vast majority of this universe is what science calls dark energy, or dark matter. It is called "dark "because we can't decode it.

The electromagnetic spectrum is 0.005 percent of what is projected to exist in this universe. And visible light, which is all that we can decode, is a small fraction of that. We are locked into this dimensional reality, which we call home, and we don't know how to get out, but we are beginning to remember.

If we come into this world of mind and we hold our connection to consciousness, then we are in this world physically—we can experience it, but we are not *of* it in terms of the point of observation. We are in this world, but we have a much larger perspective.

The Holographic Universe

The world certainly looks solid, but it can't be, because the world is made up of atoms, and as quantum physics has shown, atoms have no solidity. The reason the world appears to have solidity is because the information in the waveform base-state construct is decoded through into apparent solidity. Again, it's just the way we decode reality that gives it form.

The reason it appears solid is we live in a holographic universe. Holograms appear to be three-dimensional, but they are not, it's just the illusion of the way they are made.

Holograms are made by using a single laser beam. The beam is then split into two beams by a special lens, resulting in two laser beams that are exactly the same. One of those beams goes across the object they want to photograph. The other beam goes directly onto a photographic plate. The beam that's passed across the object goes onto that plate, and they collide, and they create a waveform (it is called an interference pattern). When you fire a laser at it, a very solid three-dimensional looking image comes up.

This is how we create our reality, it's holographic. So we are creating a holographic version of the waveform information construct in our heads. Everything "out there" then is but a holographic projection of our consciousness!

The holographic nature of reality can readily be seen in sacred geometry; so let's go back and revisit the fruit of life, beginning with a

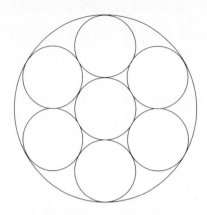

Figure 18.1. In any circle there are always seven smaller circles that will fit perfectly inside it.

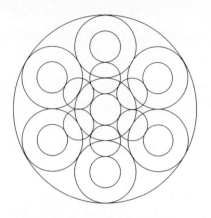

Figure 18.2. The fruit of life derived by going within the system.

Figure 18.3. The fruit of life derived by going outside the system.

simple circle. In any circle there are always seven smaller circles that will fit perfectly inside it (fig. 18.1). If you take one-half the radius of the center circle, and draw a new circle using that one half radius, and then run the circles down the three axis, you immediately come to the fruit of life (fig. 18.2). You will recall in chapter 9 how we came to the fruit of life by expanding or going outside the system (fig. 18.3); and here we see that we can also get there by going within the system. So the fruit of life is contained proportionally within every circle.

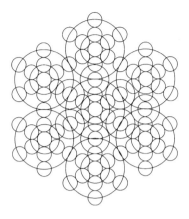

Figure 18.4.
The fruit of life connected to the
fruit of life, etc.

But it doesn't end here; you can keep going. So if we do it one more time, we end up with the thirteen circles of the fruit of life connected to thirteen circles, etc. . . . or the fruit of life connected to the fruit of life, and so on forever (figs. 18.4 and 18.5). Forever, because you can continue repeating this operation forever—there is no beginning and no end. It is holographic!

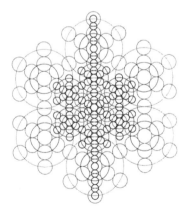

Figure 18.5.
The fruit of life connected to the
fruit of life, forever.

Every part of a hologram is a smaller version of the whole, if you cut a holographic print into four pieces, and put a laser onto one of the four pieces, you get a quarter-size version of the entire picture. Your chromosomes are geometric images and patterns that describe the entire reality; your body is a holographic image of the entire creation. The location and placement of all the star patterns in the entire universe can be derived from your body.

Since our universe is holographic, consider this as a mind-bending possibility: Our entire universe, to the furthest galaxy is no more than a closed electron in a far greater universe we can never see. And that universe is only an elementary particle in another still greater universe— and so on forever. Also, every electron in our universe is an entire miniature cosmos, containing galaxies and life and electrons. And those electrons contain an even smaller universe, and so on in both directions forever. And each piece is a smaller version of the entire universe.

So you are a hologram; a smaller version of the entire universe is contained within you. Are you now willing to consider that you're much more than you've probably considered yourself to be? You are the entire universe expressing itself at a single point!

What Does All This Mean?

We live a false identity, we think we are humans when we are consciousness, all possibility.

We worship the intellect. Yet the mind is such a low level of awareness; it was never intended to be our master, only a faithful servant of our consciousness. It took over only because we told it to.

I want to take you into that *est* training for a moment; it was a defining moment in my life, and though it was back in the day, it is still as fresh in my memory banks as if it happened yesterday.

The est training was held on consecutive Saturdays and Sundays, beginning promptly at 8:30 AM and running often until the early hours of the morning. Each day lasted however long it took for the particular day's information to be transmitted.

There was also a three-hour pre-training seminar, a Wednesday evening mid-training and a post-training meeting. My training was held in a ballroom on the second floor of the student union building on the campus of the University of California at Berkeley. There were approximately two hundred attendees.

My trainer was a guy named Hal Isen, and according to him, the fourth and final day—where the topic was the Anatomy of the Mind—WAS the training. He told us we could forget the first three days; this was it!

He asked us, "What is the mind?" And of course, he had a ready-made answer. "The mind, he said, is a linear arrangement of multi-sensory total records of successive moments of now." Essentially what that means is the mind is a stack of endless-loop recordings; complete records of past experiences that include all the senses. For some of these past events, we have conscious recall; for others, we do not.

According to *The Book of est*, the purpose or more accurately, the design function of the mind is survival; "the survival of the being and anything the being considers itself to be."[1] If the being comes to identify itself with its mind, you have ego, and at that point, the purpose of the mind becomes its own survival.

That means the purpose of the mind becomes the survival of the records, the tapes, the points of view, the decisions, the thoughts, the conclusions, and beliefs of the mind. Now the mind has a vested interest in all of these.

For the mind to survive, it tries to keep itself intact, so it will replay the same tapes, and prove itself right. It seeks agreement and tries to avoid disagreement. It wants to dominate and avoid domination; it wants to justify its points of view, conclusions, decisions, and avoid invalidation.

It wants to be right; running through it all is the unending effort of the mind to prove itself right. So, the mind is going to play one of its endless-loop survival tapes whenever it thinks it's threatened. Whenever something in the environment makes it think it's threatened, it's going to play that tape.

The next question concerned the construction of the mind. It was decided that there are actually two stacks of tapes—one necessary for survival, and the other not necessary for survival. Given that the design function of the mind is its own survival, it would consider the first stack to be more important than the second, because the mind will play the survival tape whenever it feels threatened.

In the necessary-for-survival stack, there are three different possibilities. They are labeled by *est* as a number one, number two, or a number three experience.

A number one experience is the most basic threat to survival, involving pain, impact, and relative unconsciousness. Relative unconsciousness means "anything from full unconsciousness as in sleep to the sort of semiconsciousness we experience when under extreme pain or when under a partial anesthesia."[2]

Examples from childhood could include any type of accident—falling out of a tree, a bicycle or auto accident, getting beat up, etc. But the earliest and most significant number one incident is birth!

A number two experience "is one in which the mind experiences a sudden shocking loss accompanied by strong emotion, usually negative. The easiest examples are the sudden deaths—to a child all deaths are sudden and unexpected—of a father or mother or brother."[3]

A number three is any experience that reminds the mind of an earlier number one. Since a number three is associated with a number one experience, examples from birth would include anything reminiscent of the trauma of your birth: hospitals, doctors, nurses, surgical gloves, forceps, the color of the walls in the delivery room, getting spanked, etc.

That means, for example, that upsets are never for the reason you think they are, because they are never in present time. Any upset is a past number one experience getting triggered. At that point it is a threat to survival associated with pain, impact, and relative unconsciousness, which results in the mind mechanically playing an old survival tape. From an earlier number one incident we formed certain conclusions about how we need to behave in case that situation ever presents itself again. The tape contains a complete record of how we survived. Since it worked once, the mind reasons it will work again. It is operating solely on a stimulus-response basis.

Then Hal was interested in discussing the relative sizes of the two stacks of tapes. How many records are there in the necessary-for-survival stack, and how many in the stack not necessary for survival?

He proceeded with a very logical argument, beginning at birth with an imaginary person. It was acknowledged that there would be records of this person's birth, which would be stored in the stack necessary for survival, and that there would probably be some reminders of his birth (number threes), along with maybe another number one incident. But it was also assumed that this infant would have plenty of records of experiences not necessary for survival—situations where he was playing and having a good time, where nothing came up to remind him of his birth. They, of course, would be in the other stack.

Hal asked us to consider the percentage of records that would be contained in the first stack, as compared to the records unnecessary for survival. Various estimates were tossed around by the trainees. He decided on the lowest figure of five percent—not that it was true; it was simply the lowest reasonable figure that anyone gave, and Hal was being very reasonable. So, we agreed then that after one year, five percent of this infant's experiences were recorded in the necessary-for-survival stack.

The next question involved considering what would happen in the next three years of this young person's life. It was agreed for the sake of argument that this youngster had a pretty easy life and had only one additional number one experience in these years.

So, by age four, our child has had three number ones, and a few number twos—both of which could trigger several number three incidents, given that a number three can be anything the mind associates with a one or a two.

We then considered the percentage of his experiences between ages one and four that would be ones, twos, or threes. The lowest figure given was twenty-five percent, and it became the agreed-upon percentage.

We agreed that for the next three years of this child's life the percentage of ones, twos, and threes can only increase. "If after birth there are, for example, one hundred different stimuli that can trigger ones, twos, or threes, and at the end of one year there are, say, a thousand, and after four years, forty thousand, the number of experiences that can escape being number ones, twos, or threes keeps getting smaller and smaller, and the percentage of experiences which are number ones, twos, or threes gets larger and larger."[4] This means the percentages will continue to grow at least at the rate they have, which means that by age seven, at the latest, all of our young person's experiences will be in the stack necessary for survival.

But at this point, Hal "remembers" that he has forgotten something, the illogical logic of the mind.

> After the royal number one experience of birth, the child has at least a hundred stimuli which would trigger a one, two, or three experience. But with the mind's logic of identity each of these stimuli is immediately associated in the baby's mind with everything else it is related to. The doctor's hands get associated with men's hands, get associated with men's arms, get associated with men, get associated with human beings, and so on and on. The green of the hospital wall gets associated with green leaves, with trees, then with bushes, and so on. The hospital walls get associated with all walls, with all surfaces, and so on. If we honestly look at what happens from the moment of birth onwards, we'll see that everything the baby

experiences from then on is associated with pain, threat to survival, and relative unconsciousness, that everything the baby experiences from birth on is at least a number three experience and thus all of the baby's records are in the stack "necessary for survival" and thus all of his behavior is of the mechanical stimulus-response variety.[5]

That means that from the moment of birth onward, we come under the influence of the mechanical mind. Like Pavlov's dogs, we operate solely on a stimulus-response basis. When we hear the "gong," we "salivate." From that moment on, everything is stimulus-response, we are totally under the influence of the machine mind . . . stimulus-response, stimulus-response, etc. . . .

Our mind is a marvelous instrument if used correctly. It is supposed to be a servant of consciousness; we have made it our master. As Einstein said, "The intuitive mind is a sacred gift and the rational mind is a faithful servant. We have created a society that honors the servant and has forgotten the gift."

We have reached a fork in the road; are we going to awaken to our true genius and potential as infinite consciousness? Or are we going to remain entrapped in our limited identity? The choice is between the mechanical mind which keeps you in separation, and the intuitive heart which connects you to all that is.

We live in a time that can be deeply moving, but first, you must be listening with your heart. It has been on this planet that thoughts and ideas—using the mind—has been considered to be most important. This leads us directly into polarity consciousness. When you begin to listen with your heart, you can begin to find a common language that is beyond right and wrong, good and evil.

If you identify only with your mind, and you think that you are your limited identity, then you will defend your mind to the death because that is your identity. If that is gone—if your sense of identity is gone—it's like death. Remember, the mind is a survival machine; it needs to be right.

There are two main things that a survival-based mind looks for: #1, how is it in this situation, and #2, what is the right way to do it? In a constantly changing universe, the right way continually changes—what was right before might not be right now. If you are a mind, you must look for ideas and concepts to attach yourself to in order to know the right way to proceed. This becomes cumbersome, like a bull in a china closet. When you're in a situation that requires different ideas of what's right, you can't readily shift your position.

But if you have trained your mind—if you have asked your mind to be a humble servant to your own Spirit, you will find that your Spirit knows exactly what's right in every situation and can always guide you to the right action. There is no other source of information or data that can do this for you in a consistent manner. You have everything you need, and it is all within you.

There are some design adjustments that are necessary however for adaption, so that you can access all of this information. While you are trying to access everything mentally through books, seminars, asking questions, and so on—and you think that you're getting your information from outside sources, and you think that your ideas about them are something you can hold onto for a while—it becomes very problematic. In these extremely accelerated times, you're lucky if you can hold onto a consistent picture of reality for more than a week.

You always know whatever you need to know whenever you need to know it, and you need not know any more than that. Then when Spirit has something it wants your mind to do, it gives your mind a project—minds love projects!

It isn't helpful when embarking on a spiritual path to consider your mind to be the problem. Ask your mind to help you, it loves projects. If you deny it projects, it's going to revolt and cause all sorts of problems. It's the separation you create in yourself that only creates more separation out there. Don't separate yourself into various components and shut some of them out.

Your mind exists to serve Spirit; ask it to help you to follow your spirit. Ask it to let you know whenever Spirit has a message for you. If it has some projects for you, pick them up and start doing them.

The mind was designed to be a humble servant to Spirit. The only reason it took over is because you told it to. It knew that it was too small for the job of running your entire existence. It always felt inadequate in the process and developed an ego and pride in order to compensate for not having Spirit to guide it.

Now you can have true freedom of expression. You can actually begin to let go bit by bit of ego and pride, and allow the magnificence of your own Spirit to shine through you more and more. Fulfillment can only consistently come from expressing your own Spirit. It is very difficult to find it from someone or something else trying to provide it for you.

19
Connecting to Source

Now that we've taken a look at the holographic universe, it becomes easier to see that we all exist on many levels. So even though we think of ourselves as living on Earth in a human body, we also exist multidimensionality. We have been previously connected to our higher aspects but we lost it after the Fall in Atlantis, thirteen thousand years ago. Our higher level, our Higher Self, has Higher Selves connected to Higher Selves, etc. . . . until the waveform level of universes is transcended.

If we can reconnect, we get clear guidance from within, we are present in life and are able to live in a manner that seems impossible to us now. Reconnecting is not channeling; it is just reconnecting the severed aspects of ourselves.

In Atlantis, prior to the Fall, we were in Unity. We had transpersonal memory, meaning what was available to one was available to the whole body of Atlantians. We had holographic memory, as I described in chapter 6. We were In Unity consciousness, and as such, we were heart-based. In Unity there is no separation; you see yourself in everyone, so to mistreat another is to mistreat yourself.

We are now clearly in separation; we look out at a world that isn't us. We don't experience our intimate connection with our biosphere, we stand apart from it and try to control it. We don't experience oneness with each other either, we are often times more interested in getting ahead than in getting along.

What happened in the Fall is we moved from our hearts (only knows Unity) to our minds (only knows separation). We became our minds and thus became separated from our true nature. The mind is polarized, we see good and evil, and we are constantly judging. And as I stated in chapter 12, there is no such thing as a polarity. For every polarity you can think of, there is always that third, middle, or unifying component. For example; for black and white, there is gray, for hot and cold, there is warm. You can go on and on with that.

In fact, time is also in threeness (past, present, future); and so is space (macrocosm, the space in which we live, microcosm); so are mathematical sequences (if you have three consecutive numbers, you can easily figure out the entire sequence).

We exist in threeness too; Higher Self, Middle Self, and Lower Self. When we are perceiving Reality from our mind, we are living from our Middle Self. We are cut off from Source and only see the illusion of polarity.

We must first go down in the process of reconnecting our severed aspects, so we can see and live wholeness again. We must first connect with our Lower Self. The Lower Self is our subconscious mind; it is a child, about two to six years old. The Lower Self is not just our own subconscious mind; it is the subconscious mind of the entire planet with which we are connecting. We have lost this bond, we have severed our relationship with the Earth, and we very much need to reconnect.

Once you have connected with the Lower Self, you can make connection with the Higher Self. But you can't force it—it will happen only when the Lower Self knows you are ready, and it will then arrange it.

Did you see the movie *Contact* with Jodie Foster? As the movie began, we were introduced to Jodie's character, Ellie, when she was a child. She had that total sense of awe and wonder that children have.

For her it was the vastness of the Universe, and of considering the possibility of extraterrestrial life.

As an adult, she still had the total commitment, but it became mental. She was a scientist (she became her mind). She didn't believe in Oneness, unless she could prove it through science. Her interdimensional experience reconnected her with her inner child, and then some!

Now I would like to take you back to the magic of your childhood. Do you remember when you were maybe four or five years old? Perhaps you were looking up at the sky on a beautiful star filled night. Do you remember the sense of awe and wonder as you gazed at the Belt of Orion, the Big Dipper, or in fact, any other constellation. You didn't perhaps have the words for it, but you didn't need them, you *felt it* in every cell of your body; your deep connection to all that is.

For Jodie Foster's character, Ellie, it took a fantastic interdimensional experience for her to rediscover the pure presence of her inner child. You probably will never have an experience like that, but then again, you might! However, it doesn't take some incredible adventure to reunite ourselves with the innocence and with the sense of awe, wonder, joy, and excitement of our youth. It is still there; we just have to rediscover it.

I will submit that existence itself, just the fact that you exist to be reading this, is an incredible miracle. Is it your experience that your life, just the fact that you exist, is a miracle of unlimited proportions? Now, I realize that you probably are not in a continual state of blissing out on the miracle of your existence. If not, why not? Could it be because we are in a constant state of judgment; that our mind is continually comparing our situation at hand with some imaginary standard of how it "should be"; and if it doesn't measure up to that standard, we "make it wrong?"

Perhaps another part of it is the adult trap we've fallen into of looking to find fulfillment from outside sources, from our circumstances. You know if only you had a better job, or if only you had the right relationship; then your life would turn out! You will never find contentment much less your true Self by looking outside. You find it only by going inside. You begin by acknowledging your wholeness and bringing that into your life situations. Use that as a place to come from.

Children are not waiting to see if they can find gratification from their circumstances, they bring their sense of awe and wonder, and innocence, into whatever they are doing. And then whatever they are doing becomes a place in which to experience and express the context of wholeness that they have naturally created.

We can do that too! We have to rediscover our childlike innocence. Children are not in their minds; they are not living life conceptually. They are not going through the motions. They are experiencing life directly, they are present, and they are functioning naturally from their higher centers (Higher Self), where awe, wonder, adventure, joy, and peace exist in infinite supply.

They only know Unity; we teach them separation. It takes twelve years or so to complete the job. Consider that anyone under the age of twelve is your teacher—animals too! I live with a Zen Master, a cat named Alfie. He teaches me everything I need to know about presence, joyfulness, and innocence too!

Experiencing life directly is a function of letting go of some of our adult sophistication, and rediscovering our childlike nature. And it is only available in the present moment. As you observe children, you will notice that they are present, they are in the experience of life!

What do I mean by being present? Well look at yourself, are you ever in the present moment, or you in the past or future? How about when you are at work? How often are you in the moment with regard to what you are doing? Are you more often dwelling on the past, or some imagined future; like 5 p.m. and what you'll be doing after work, or your plans for the weekend, or how much better it will be if only you could get a promotion?

How about when you are driving home after work? Most people are on autopilot; they are busy thinking about dinner, or how the work day was, or their vacation plans. I think you know what I mean.

This is not to say that you are never present, you are. Whenever you are truly enjoying yourself, and time seems to fly by very quickly, you are in the present moment. You have stepped out of time, and lost yourself in the moment.

What I have found is that for most people, the present moment is too painful, so they don't spend much time there. Yet it not only is the only place where experience exists, it is also the doorway to your true nature, your Higher Self.

In the chapters that follow, I will show you exactly how to be present, and that it not only is not painful; it is in fact your source for unlimited amounts of joy, inner peace, unconditional love, all of which lies beyond your mind, and all of which is contained within you. It is your true nature; it is your connection to Source through your Higher Self!

Forgiveness

20

In Order to Ascend,
We Must First Descend

In the previous chapter I said we must first go down in the process of reconnecting our severed aspects, so we can see and live wholeness again. We must first connect with our Lower Self and rediscover our childlike innocence. In so doing, we reestablish presence; and it is only in the present moment that the doorway to our Higher Self is open.

We came into this world from another dimension, and as such, we only knew Unity. We are taught separation, and it begins at the moment of birth. We become totally identified with our minds (our ego or false self) at the expense of our true nature as infinite consciousness having a human experience.

Mind-based creation (living from the Middle Self) can give you most anything you want. However, it will also give you exactly what you don't want. Because of its polarized nature, the mind is in a continual state of judgment; it is busy comparing reality to its imagined ideal standard of how the person or the situation at hand "should be." We are so busy insisting that reality should be different from what it is, that we don't let ourselves have the experience of benefit and pleasure that's available in the experience that we're actually having. And that idealized standard is rooted in our conditioned past; so, the mind sees the present through the eyes of the past. It is therefore never in the present moment; presence can only be experienced, and never conceptualized. By the time it takes to conceptualize, the experience is gone and only its memory remains.

Living from our mind then is living life conceptually from our conditioned past, and it reduces us to "just going through the motions." There's no real satisfaction because there's no real presence; there's no Self from which to come from to consciously create our reality.

Reconnecting our severed aspects and fully reclaiming our child-like innocence is code for the healing work that needs to happen, and that begins and ends by healing our most fundamental relationship—that of our parents.

Before diving in, I would like to take a moment to acknowledge the many teachers who have helped me in understanding and unraveling my own incompletions with my parents. Included on my list are Leonard Orr, Diane Hintermann, Bill Chappelle, Phil Laut. Lia Holtzman, Catherine Ponder, and Werner Erhard. Werner is at the head of my list; in addition to the *est* training, I took a daylong seminar from him on healing our relationship with our parents. What I learned from him sticks with me to this very day.

Now with that being said, let's begin by returning to our early childhood for a moment and see what happens.

In revisiting our discussion of the mind in chapter 18 (all of which came from Werner Erhard) and recalling that the purpose of the mind once the Self considers itself to be its mind is survival of the mind itself. For an infant, life is a very heavy survival game; there's the business of getting fed and getting your diapers changed, and getting someone to pay attention to you, and being reassured that there's someone around to take care of you. You are totally dependent upon your parents for your survival.

Then there comes a place and time in your life where your parents withdraw their support; and that is enormously important to you. And at some point in everyone's life, parents do withdraw their support.

I will give you an example from my own life. I was three years old and I was "being good," meaning I was already learning that love and approval was not unconditional—it was dependent upon my behaving in certain approved ways. It must be earned, and there was punishment for improper behavior.

I was with my mother in the kitchen, "helping" her make a cake. Since it was a marble cake I decided the best way to help was to fetch my bag of marbles and offer them to her for the cake—after all it was a marble cake. As I began to take the marbles out of the bag, somehow, they all spilled onto the floor, and that was upsetting to me. What I needed at that moment was love and understanding and reassurance. I needed to be told that I wasn't being bad, that it was just an accident and that everything was okay—and in fact, she would help me pick them up. What I got instead was being sternly told by my mother to pick them up, and I went berserk. I threw a tantrum because I saw that as punishment, and I couldn't understand why I was being punished for being good. After all I was helping her make the cake, so why was I being punished? Then my father got into the act and what began as a fun experience had by now become a true nightmare.

Prior to that and even though I had learned from my father that there were consequences for "improper" behavior, I knew I could count on my mother for support no matter what. That all changed in a moment and I was devastated; I felt helpless, insecure, alone, and abandoned. I was left with a gaping internal wound that shaped my relationship with my parents and was not healed for many years.

Let me take a moment to share with you a process that I have used on a repetitive basis that helped me greatly in resolving this huge incompletion in my life. It will help you too.

Close your eyes and visualize the child you were as clearly as possible, at any age up to five or six years. Give the child a childhood name that you were called, an endearment, or a nickname.

Imagine this child at any age up to five, sometimes as an infant, sometimes older, but not much older than five. Imagine the child always has the ability to talk.

Be aware of what the child inside of you feels and needs. This child needs to feel loved and accepted, needs to be reassured and made to feel secure, needs to be relieved of guilt.

Visualize the child five feet away, tell the child what you wanted to hear when you were little. Express love and reassurance to the child in order to take away feelings of insecurity and anger and guilt. Give the little child in front of you the touching, hugging, and holding you always wanted.

Now repeat the following expressions to the child. "I love you; I love you more than anybody else. You're the most wonderful child in the world. I will always take care of you. I love you and I'll never leave you."

See yourself exchange gifts with the child as a token of your eternal love.

This withdrawal of support is a big deal for many of us, so let's take a deeper look, beginning with another example. Suppose that you're a four or five year old boy. Dad sees you crying; what you need is for Dad to let you know it's all right. You need understanding, love, and compassion. Instead, because Dad wants you to grow up to be a man; maybe he wants you to be a football player, or a Marine. So instead of support and reassurance, you get "What's wrong with you? Don't be a sissy, don't be a crybaby, be a man!"

You just got invalidated for your feelings of insecurity, sadness, fear, and loneliness. And you absolutely need Dad's support and approval; your survival is dependent upon it! After all he takes care of all your needs—food, shelter, clothing, and so on. So, at some place there's a shift in our relationship with our parents from one of being supported to one of not knowing if you can count on being supported; and that's something for you to be aware of. Prior to that your experience was that your parents would support you no matter what, and when you were a particular age that shifted. That really rocked your world, and you've been acting out the upset with your parents ever since; and in ways that you are probably not conscious-mind aware of. This "acting out" either takes the form of rebelling against them, or by having concluded that there's something wrong with you for feeling what you were feeling. If you did conclude the latter, in order to get by in life, in order to survive, you learned to control your emotions, suppress your feelings of guilt, resentment, and other hurtful feelings and develop an act in order to get the love and approval that you so desperately need. All this gets reinforced in school and in religious upbringing with other authority figures.

But whether you rebelled or conformed, if you've never forgiven them, you are completely at the effect of your parents. And that means that your entire function as a person is directly related to your incompletion with them.

Then as Werner says, the other part of the survival game is that your mind just records the pictures—it just records the incident. And then it brings those pictures or incidents out whenever something anywhere near or like them, comes up. If you've got a picture of where you lost and somebody else won, and that gets reactivated, you will now play out the winner. When that picture gets reactivated you don't play yourself; you get the other person to play you and you play the winner.

And if you look at your childhood, you see that your parents won all the time; you didn't ever get to win. You didn't ever get to be right as a kid; they were always right. And they were bigger than you were too. They were tougher and smarter; they had more money and they knew more stuff. What happens for a lot of us is that in order to survive, we have to be one of our parents.

I became my father whom I resisted all the way in my never-ending attempt to prove to him that he couldn't dominate me, and I didn't have to do it his way. Once I decided that the only way I could survive was to be my father, I had to be better at being him than he was. I did this by rebelling against him and throwing his commands right back at him in my ongoing attempt to prove that he had no hold on me. So maybe you too decided you had to become one of your parents in order to survive. If so, it can be very useful in the completion process to discover if you're stuck in either your mother's or father's identity.

All this entanglement results in a huge reservoir of unresolved emotions; and those feelings don't go away, they live in your body as stuck energy. Then because you want to feel good, you do everything you can to distance yourself from those unpleasant feelings; you suppress them. You at best, only find temporary relief; you are only masking the symptoms. And you can be certain that these suppressed feeling will get reactivated, and usually at the most inopportune times. Around and around you go; you are a puppet on a string!

So, suppression and controlling our emotions is something we first learned from our parents that continued in school and religious upbringing as the only way of dealing with our feelings that parents and other authority figures in our life didn't like. The parent abuses the child to force him or her to suppress feelings that are not okay with the parent. This is emotional abuse; it is behavior that adults use to control their children so that the adults feel better. The abuse usually continues until the child gives up; surrenders his or her divinity in the name of following instructions and suppresses his or her feelings. The spirit of the inner child (Lower Self) is then broken as we become our minds.

In view of this, you may be wondering why you would want to continue your relationship with your parents. Maybe you still fear your father for the emotional and perhaps even physical abuse that he dished out to you as a child, and you still feel quite uncomfortable in his presence. It has therefore seemed like the best way to deal with him is to avoid him as much as possible, and possibly even declare a divorce.

Let me state unequivocally that what we're after here is the creation of a context, a space in which to be and live, called completing your relationship with your parents, the source of your relationships. And as Werner correctly points out, the bottom line is that we are complete in our relationship with our parents and then over the top of it is all this stuff. Then we get tied into the stuff; you know we get tied into the problems, we get tied into the sadness, the resentment, the embarrassment, or whatever it that we've got going on with our parents. And then we lose touch with the context of that stuff; and the context is that you and your parents are related, absolutely totally completely related. You can't be any more related than that; that's the source of relationship. That's where you as an individual come from. So what we're after is a recovery of that context you have with your parents; and that context is that you are complete in your relationship with your parents.

Completing your relationship with your father will not make him less fearsome; it will simply get you off the string as a puppet of your fear of your father. It's not that he's fearsome, it's that you're afraid of your father. He may well be fearsome and you might handle it by not getting too close to him. The other side of it is that your father hasn't been able to complete his relationship with you; he's got a message that you've never gotten, because you're too busy being frightened. It could be that he's not really fearsome and when you give up being attached to being afraid of him, you may see that your father is really different from the way you thought he was. He may be very stuck on something, rather than fearsome. Remember, he too had parents from whom he learned to relate.

Our parents disapproved of us because they were disapproved of by their parents, because they were disapproved of by their parents, etc. Abuse and suppression then, is passed on from one generation to another. And unless you break the pattern, you too will pass it on to your children.

So this is not about blame; your parents did the best they could, given what they had to work with. Rather it is about taking responsibility to work on yourself; that is if you are truly interested in finding a permanent solution to the stress and disharmony in your life.

As you can see then, we have been well conditioned to control our emotions and what we're allowed to feel. We become good actors and actresses—our lives then become an act, a cover, a camouflage to hide our feelings and to appear to the outside world as though we are "normal."

By completing your relationship with your parents, I'm not suggesting that you either continue or end your relationship with them. I'm simply pointing to the importance of being complete with them. Since parents are your fundamental relationship, anything unresolved between you will come up in your other relationships. In fact, your life is about your incompletion with your parents—you will find yourself being shaped by, dominated by, and limited by this incompletion.

If you've been working on the other relationships in your life and your relationship with your parents is incomplete, you're not really working on the other relationships; you only think you are. Until you are complete, you will find yourself creating "substitute parents," where you unconsciously tend to recreate their personalities as closely as possible in your other relationships. That means you will probably discover at some point that you married your mother or father, or your boss is your father. So in order to make it with your mate or at work, you first must complete your relationship with your parents.

The point is then, that until you complete your relationship with your parents, your life is very much about your relationship with your parents unconsciously. Those internal commands are made up of the stuff that's incomplete for you with your parents until you do in fact experience being complete with them. No matter what the circumstances are, you can have a complete relationship with them.

A thing is complete when you can let it be the way it is. Anything in which you're not able to let be, you are therefore, incomplete with. The way you complete your relationship with your parents is by allowing them to be exactly the way they are. When you are able to allow them to be exactly as they are, and to be responsible for your experience of them being that way; when you're able to allow them to be, then your relationship with them is complete. And anything which you can allow to be, will allow you to be.

It begins with your willingness; without that, there is no hope. Without this willingness, you are resigning yourself to a life of going through the motions. You do get a payoff though; you get to be right (remember, the mind has an insatiable need to be right)—you get to pay them back for what they did to you! But at what cost; is it really worth it? And will you ever get even enough?

Even if they did emotionally and perhaps even physically abuse you, it's a total waste of time to continue to blame them for it. You will only stay stuck for as long as you do. So the question is, are you willing to take responsibility for your relationship with your parents, or do you have to blame them for it? Are you willing to be responsible for your own feelings—if you felt unloved, are you willing to be the author of feeling unloved? Are you willing to shift from being victimized to being at cause in the matter? All those upsets and petty bothers, are you willing to come out of the space in which you are the creator of all of that? If so, your realization will shift from "that's the way they relate to me" to "that's the way I relate to them."

And if you always judged them and always resisted being like them because you didn't want to be like them—you wanted to be like yourself—you will by shifting from effect to cause in the matter, get that the way they are is okay and it's okay to be like them. And underneath it all, you love them very much. If not wanting to be like your parents is the command you give yourself, that precludes any possibility of being yourself, because what you're doing is not being your parents. And that isn't any different than being your parents. There's no difference because whatever you're resisting, is determining your beingness. You resolve it by being willing to be exactly like your parents, and that frees you to be yourself. And the one thing you don't have to work at is being yourself.

Let's suppose that you really are incomplete with your parents and then after reading this chapter you say, "No my relationship is complete." And then in your attempt to live as though you really are complete with them, you go into denial of the unresolved problems and upsets. What you're doing then is smearing completion on top of incompletion; and all you get is incompletion magnified or expanded. So that isn't the secret; I'm not telling you to invalidate or try to oppose or resist your point of view that your relationship with your parents is incomplete. If that's what's true for you, that's fine, that's alright. Don't plaster that over with positive thinking!

What I'm saying is, take a step up in your abstraction and create a context in which the point of view that you are incomplete with your parents is simply another content. And the context is you are complete in your relationship with your parents. This reduces all the resentment and anger and sadness and all the stuff that you've been holding onto to mere content. You are no longer at the effect of it because you now have a container in which to hold it; and it now becomes possible to integrate it. It is essential that you allow yourself honest awareness of the feelings you suppressed as a child and resolve them, which you can now do by applying the Five-Step Harmonizing Method. I will show you how to do that in the Breath Alchemy chapter.

At the bottom of all relationships is the fact that parents absolutely love their children, and children absolutely love their parents; and there's never been an exception to that. Then layered on top of that love is a tremendous amount of resistance, lies, irritation, resentment, and hatred—all that stuff. And then having to hide that negative stuff—by suppressing it and keeping it unconscious and out of the way—barriers people against the experience of loving each other. It's just having to hide that stuff and not allowing it to be. Or if you do allow its expression, you become the effect of it. If you will take responsibility for the barriers, the problems—if you will accept them and allow them to be, what you'll discover that what's underneath them is the experience of loving your parents and being loved by your parents. And I mean love and all that word implies and conveys. It's a really very powerful expression at that level—powerful enough to restore your childlike innocence and open the channel to your Higher Self!

Forgiveness

Definitions of Forgiveness:

> To cease to bear resentment against
> To give up resentment against or the desire to punish
> To stop being angry with
> To pardon, to overlook
> To give up all claim to punish, or exact penalty for
> To give up all claim to punishment as well as any resentment or vengeful feeling
> "Give for" or to "replace" the ill feeling, to gain a sense of peace and harmony
> To give love for yourself

Take a few moments and decide the following. You can either write down your responses, or you can just do it in your head. With your mother, what are you willing to forgive her for? With your mother, what are you willing to accept her forgiveness for?

Now do the same with your father. With your father, what are you willing to forgive him for? With your father, what are you willing to accept forgiveness for?

The purpose of this little exercise is not for you to become willing to forgive your parents, but for you to discover exactly what you are willing to forgive them for, and exactly what you're willing to accept their forgiveness for. If you discover that you are only willing to forgive them for certain things and not for others, that's valuable. If you really get in touch with the fact that you're not willing to forgive them for something, suddenly it begins to take care of itself. If you would like some help with this, contact me at my website BobFrissell.com/contact and I will offer you some additional suggestions.

Conscious Mind - 10%

Subconscious Mind - 90%

21
Creating Reality from Within

In the Infinite, there is no time, no place, no divisions, no sense of "us" and "them." Infinite consciousness *is* the only truth. The heart only knows Unity and Oneness and is thus able to take us beyond the illusory walls of limited perception.

The right brain is our connection to the *All That Is*; it is holistic, intuitive, experiential, and knows only momentary time, *the eternal now*. It is able to intuit the Oneness, like a four- or five-year-old, looking up at the night sky. The left brain sees dots and does not connect them. It is logical, and is locked into linear time, so it is never in the present moment.

We need both however, and we need them harmoniously working together. The mind looks out at the reality and sees separation; it does not therefore believe in Oneness. It must be shown Unity in a series of logical steps as I have done with you by showing you the holographic universe through sacred geometry and by showing you how the reality that we think we see "out there" is really a holographic projection of our consciousness. When the mind (left brain) truly sees Unity, the corpus callosum (neural fibers connecting the two hemispheres) opens up, communication happens, integration takes place, a relaxation occurs and we become whole again. Because mind is now serving consciousness instead of serving itself, we can replace beliefs with intuitive knowing and learn to follow Spirit. The doorway to the heart has thus been opened.

You are the Self-expression of the Infinite; you are the entire universe expressing itself at a single point. Within you is the same presence, knowledge, wisdom, and power through which the entire universe was created.

Your Higher Self within you is your connection to Source, and as such, is always in alignment with the infinite perfection of your true nature. It is always bringing forth thoughts of perfection . . . perfect health, perfect order, the perfect solution to every situation, total abundance, and total fulfillment in every way.

But you must claim it in order for it to be so in your reality. You must first be willing to let go of your adult-like sophistication and reclaim your sense of awe and wonder and excitement—your childlike innocence, as detailed in the previous chapter. You must also change your identity. You are not a human being having a spiritual experience. You are a spiritual being having a human experience!!

Until you make this declaration, your spiritual essence must flow through the barriers that keep you from the experience and expression of your infinite perfection. It must flow through a consciousness of imperfection; all your limiting beliefs. It will change its purpose, objective, and destination according to the tone of your consciousness.

If it passes through a belief in scarcity, that you do not have enough money to make ends meet, your unlimited potential becomes limited, and it will move Heaven and Earth to make sure that there is an insufficiency of money to meet your needs.

The same holds true for health. In the mind of your Higher Self is the idea for your perfect body as well as perfect health. This idea of perfection is forever expressing itself, sending forth the perfect image into every cell and organ of your body.

But again, the creative energy must flow through your consciousness, and if your mental atmosphere is charged with a belief in sickness, disease, and injury, then you will attract this to you.

Declaring your Divine Essence

So, you must claim the truth of your divine essence, that you are a spiritual being. This creates a context that reduces all those barriers, the limiting beliefs, to mere content. In so doing, you open the doorway and allow your connection to Source to come through you.

Then you learn to create a "gap between your thoughts," where your beliefs no longer affect your reality. Doing so aligns you with universal principles and matches your energy with the energies directly from the field of all-possibility—those high frequency energies of love, kindness, inspiration, passion, joy, and so on.

Okay, so how do you do this? You do it by discovering as I did, three ideas that are so powerful, that embracing them will completely transform your life, and I do mean completely! I first heard of them as the three notions of *est* way back in the day when I took the two-weekend *est* training. Since they are so all-encompassing, I simply call them Universal Principles. By the way, these three ideas or notions can be referenced in *The Book of est*.

Let's begin with the idea that you are perfect, but there are barriers preventing you from experiencing and expressing your perfection. That is the *Perfection Principle*.

And the second has to do with one of those barriers blocking you from feeling and expressing your divine perfection. I call it the *Resistance Principle*. It says that resistance leads to persistence.

As I said in *Something in This Book Is True,* "If you try to resist or change something, it will become more solid. The only way to get rid of something is just to let it be. This doesn't mean to ignore it. Ignoring is actually a form of resistance. To ignore stress, anxiety, or anger is one of the reactive minds ways to try to eliminate it. To let something be means to observe it and stay in touch with it, but make no effort to change it."

The effort to control or change something absolutely ensures its persistence. If you're angry and try to change it, your anger will persist.

If you're feeling tense and try to relax, you'll continue to be tense. If you have a headache and try to change it, your head will continue to ache as long as you're trying to get rid of it. This is also true for feelings of fear, sadness, guilt, shame, stress, anxiety, impatience, boredom, depression, bitterness, and so on.

So, this is not about change, it is about whether something persists or not. If you try to change tenseness, you may change the form of the tenseness, but the tenseness will still persist. It will not disappear; the substance will remain the same.

The third law is what I call the *Harmonizing Principle*. It says that the re-creation of an experience makes the experience disappear. Re-experiencing to completion disappears it. To re-create an experience, you get totally in touch with it, you rebuild it element by element until it is entirely restructured and the paradox is, it disappears.

So, using the example of tenseness, you first have to get in touch with the elements of tension. You get in touch with its exact location, its size, shape, color, and especially, you get in touch with what it feels like in your body.

"Tenseness" is a word that people use to try to describe a certain kind of experience. People don't know their experience, because they are living in their mind, where experience is conceptually derived. That means they are living in the realm of nonexperience, so they don't really know any of the elements of tenseness. Now just what do I mean by that?

Okay, how about if you go into a restaurant, sit down and look at the menu to help you decide what you want to order. Some restaurants even have fancy pictures of the different meals that they serve. Now you know that the menu and the fancy picture of the meal is not the actual meal; it is a conceptual representation of the meal.

And the same is true for tenseness. There is your mind's narrative of the experience; and that is only a conceptual representation of tenseness—not the real thing itself (just like the menu). The actual experience of tenseness lives in your body as a sensation, a feeling.

Now you would never confuse the menu for the meal in the restaurant, yet we do it all the time with our unresolved emotions. The mind has a story about how bad it is, and this story is but a repetitive recording—you will become increasingly aware of this as you practice watching your thoughts as an impartial observer. The story is only the conceptual representation of the actual thing, yet we consistently confuse it for the real thing itself. We are eating the menu! The actual thing that we call tenseness lives in your body in the form of bodily sensations which is stuck, stagnant energy that doesn't feel good because you're making it wrong.

So, in order to get beyond quick fixes and symptom masking, you have to get to and eliminate the root-cause of the problem. You have to let go of resistance and apply the Harmonizing Principle to the actual thing itself, which is the feelings that you have been resisting.

If you will stop making war with the feeling of tenseness, stop trying to change it, and make peace with it by just being with it, observing it, and feeling it in a context of acceptance; you would in fact re-create the experience of tenseness and it would disappear.

Trying to change an experience makes it persist. Re-creating an experience—accepting it, being with it, observing it, and feeling it—makes it disappear.

How to Create Reality from Your Higher Self

We all have things that we say we want in life, and the truth is that only ten percent of us get the things that we truly want. The reason is that below the words that you're speaking, there are five things going on. And that's creating a start-stop pattern, where you're saying that you want something, yet you're putting your foot on the accelerator and on the brake at the same time. And you sputter to a stop, and most people can't break that pattern.

So underneath the surface, what's going on is x, y, and z. And until x, y, and z are addressed, you're going to find yourself in that start-stop pattern for your whole life—and you may spend more time in stop than you do in go.

The three things that are stopping you from getting what you want are thoughts, beliefs, and emotions. And where thoughts, beliefs, and emotions show up are in your relationship to yourself, your relationship to others, and your relationship to the world.

Some people do really well with their relationship to themselves and to others, but where they break down is in their relationship to the world.

Others are fine in their relationship to others and to the world, but where they break down is in their relationship to themselves and their inner dialog—they're at war with themselves every day.

And some people are stuck in all three.

So far, I have given you three Universal Principles, and I'm about to give you another one. These are universal laws—they work unerringly—and just like man-made laws, ignorance of the law is no excuse.

What you believe about life now is what controls your life, the *Law of Cause and Effect*.

This principle says you always get what you want; there are no exceptions.

Wanting is defined: conscious and subconscious. Most of what controls us is on a subconscious level. Wants are a result of beliefs. Many of our beliefs are acquired under circumstances that cause us to suppress them immediately after we adopt them.

Therefore, we don't remember where they came from—in fact, we aren't even consciously aware of many of them.

So, most of our beliefs, the operating principles in our lives, are on a less than conscious level, and we have thousands of these beliefs. They become the unwitting filter through which we see life; it's sort of like putting on a pair of dark glasses and then forgetting that you are wearing them. Yet they are creating our life moment by moment. Why?

Because the brain decodes reality through these beliefs, and it becomes a self-fulfilling reality.

What you create you create because you want. You want it because you believe it. This is a result of a belief that this is necessary in your life.

We are picking up information from the cosmos and decoding it into our experienced reality. What we do and do *not* decode from it is dictated by how our decoding system works, and that is dictated by our sense of perception. And our belief systems come from our sense of perception.

When we are in mind-based separation consciousness, as we have been for the past thirteen thousand years, we do not connect the dots to see wholeness. We see a polarized world (good and bad), this has become the unwitting filter through which we decode reality. We judge everyone and everything. When you judge anything, from an energy standpoint, you lock the energy up tight, and where you lock it is always in the same place, in your physical body.

Our beliefs then, are based in fear and limitation, and survival (how to survive in an unsafe world), so we create separation. These beliefs are creating our life moment by moment, because the brain decodes reality through these beliefs, and it becomes a self-fulfilling reality.

Research has shown that when we have rigid beliefs, the brain fires off neurons in a specific network which decodes reality in line with what these neurons are firing. So, it edits its reality and creates a self-fulfilling prophecy; it keeps seeing the world according to its belief, despite receiving information that challenges that belief. It filters out that input in order to protect its belief. Then when you change your belief system, the neurons start firing in a different way and you begin to see the world differently.

We *are* creating everything that happens in our lives, both the good and the bad—there are no exceptions. It's easy to claim the good; and in many cases it's much more difficult with the less-than-good. You might wonder if it's a bad job or a major upset with a friend, "Why would I create that in my life? I certainly don't want *that*!!" And furthermore, there are probably many situations in your life where you could easily get agreement from your family, friends, and colleagues, that it really was the other person who did it! But that will only serve to keep you stuck.

There is one simple way to use the Law of Cause and Effect, and it's the only way to use it successfully, and that is backward. If you see what you have created in your life, you know you wanted it. Done that way, you will never fool yourself. If *anything* occurs in your life, you know you created it. Not *only* did you create it, but you want it to be precisely the way it is.

Many people have difficulty with this, and seem to be forever looking for exceptions to this rule of law; the situations that seem so outlandish that it's hard to believe that they have created those things in their lives, and they ask, "How could this possibly be a result of cause and effect?" Well, it is!

What's fascinating about it is this: If you were given one wish, and you were told you could have as much time as you wanted to decide one thing, one power that you could have. After you have read all of the literature and consulted with the most brilliant people and given as much time as you required and you finally chose the one thing that could be the greatest thing that you could possibly have; what you would ask for without question, is the Law of Cause and Effect.

What does it say? It says, you can have *anything* you want, and in fact, you get it!! Our task then, is to learn how to use this law consciously. Everything that you experience in your life is a projection outside of you of your state of consciousness, that's all you ever experience. So look at life as though you were looking at a mirror; *everything* is a direct reflection back to you of what you are asking for.

We do create everything that happens in our lives, and it's wonderful that it is that way, because as you get in touch with universal principle, as you get in touch with that part of you that is real, you will experience such magnificence that it is beyond the ability of the human mind to describe it.

We tend to believe that we are who we project ourselves to be, and such good actors and actresses are we, that we do a good job in convincing each other that we are how we project ourselves. So we get locked into this crazy situation of acting certain ways that are much less than perfect, and much less than who we really are, and getting the reinforcement from others who respond to us the way we project ourselves.

If we can let go of beliefs and come from a state of knowing—consciousness rather than mind-based beliefs—then we open up our range of understanding well beyond anything that is accessible from beliefs. When we open our mind to consciousness, everything changes. The trick then is to break this spell, we have to get in touch with the fact that we are not our beliefs. We are not our thoughts, we are not how we act, and other people aren't either.

1. Recognizing that you always get what you want, you take responsibility for the fact that you created the situation even though you may not be conscious-mind aware that you did. We know from the principle involved that you did indeed create it; and principles are unfailing. Until you do own it, you will just keep experiencing it; and it will only increase in intensity for the simple reason that you're putting your energy into it, and by focusing on it, it expands.

2. Because we have falsely identified ourselves with our polarized mind, we are in a continual state of judgment. Make-wrong or judgment in every case, is a comparison. It means you are comparing what you are actually experiencing to an imaginary standard of how you think it should be, or how you wish it was, or something like that. In truth however, there is no such thing as right or wrong, good or bad in the physical universe. Everything that happens is just another event that you singly or in group created.

When you judge anything from an energy standpoint, you lock the energy up tight, and where you lock it is always in the same place, in your physical body. Judge anything, judge anybody, and what you end up doing is creating extraordinary energy blocks in your body which you carry around as pain.

So it's the judgment, the make-wrong that holds the unwanted condition in place. Why? Because you are locking horns with the Resistance Principle, and it says resistance equals persistence. The very act of judging is so counterproductive that you cannot get on with your life when you do it; it is a total block. The only solution then, is making peace with the judgment.

3. This is a critical step; how would you like to feel? Since we all want to feel good, the answer is pretty obvious. And to do it you have to make friends with the Harmonizing Principle.

There are actually two ways to use the Harmonizing Principle; one way is to use it to integrate the situation you have the negative feelings about, and the other way is to integrate the unpleasant feelings themselves into your sense of well-being. Either one of these approaches will work. However, there is a potential problem using the first method; and it is somewhat analogous to letting the fox guard the hen house. The problem is that the mind you are trying to change is always the same mind that is directing the process. It is quite capable of tricking you into falsely believing that you've resolved the situation, and you're left with positive thinking. That only creates an even thicker barrier between your conscious mind and your subconscious mind by plastering over the unresolved feelings. Therefore it is extremely useful to know that if you make peace with the feeling by shifting the context in which you are holding it, the mental context (the situation you were making wrong) will also shift.

Breath Alchemy, which I will explain in detail in the next chapter, is the practical application of the Harmonizing Principle, breaking it down into a step-by-step process (known as the Five-Step Harmonizing Method) that you can use effectively and efficiently to resolve any unwanted condition. It does so by going well beyond symptom masking and quick-fixes and harmonizes the distortion in the base-state; it goes to the root-cause of the problem and resolves it by transmuting the energy at the feeling level, so you end up feeling good again.

Transforming or transmuting at the feeling level is the process known as integration; it creates this "gap" between your thoughts and allows the pure presence of Source energy to come through you; namely joyfulness, inner peace, creativity, and unconditional love.

This re-creation of what has habitually been held as an unwanted experience then, is what produces the result in my Breath Alchemy Technique. It includes the realization that thoughts and feelings are the same thing perceived through two different internal senses, so every thought has a corresponding pattern of energy, a feeling in the body. You practice this principle by learning to impartially watch your thoughts and to feel their corresponding patterns of energy (*feelings*) in your body.

Doing so gives you perfect access to the content of your subconscious mind, to the core of limiting beliefs that are unerringly producing results based on fear and limitation. It gives you access to the energetic (feeling level) component of your subconscious mind, If your conscious mind really knew what was going to solve your problems, the problems would be solved. Part of the nature of the conscious mind is that it doesn't know what is in the subconscious mind. Your body does know; and when you integrate at the feeling level, you have simultaneously integrated at all levels and throughout all of your senses.

Re-creating your thoughts and feelings enables you to disappear them; it allows you to transmute them into the light and truth of your Higher Self. This leaves you with space—space from which to create from your Higher Self. It is the same principle as turning on a light in a previously completely darkened room. Darkness cannot survive in the presence of light; the light transmutes the darkness into itself.

So, you become an impartial witness by listening to the voice in your head as often as you can. You want to begin to notice the repetitive thought patterns; the old endless-loop recordings that have been playing in your head perhaps for many years. As you watch and listen to your thoughts, you will begin to feel a conscious presence—your deeper self underneath the thought. The thought then loses its power and quickly subsides because you are re-creating it instead of energizing it through identification with it. The same holds true when you are feeling with detailed awareness, by giving your total attention to whatever you are feeling most prominently in the moment.

The Mind and Upset Tapes

Whenever your mind wants to take you into an upset; there is the make-wrong thought, and there is its corresponding pattern of energy (the feeling) in your body. You will discover that you can catch the feeling component by choosing to stay present. You can see that you have a choice; you can get sucked into your reactive mind (concept), or you can tune into the feeling in your body (experience).

This creates the possibility for both an AH HA and a HA HA experience.

You can literally watch your mind in its survival mode, and how it wants to take you into the story (endless-loop recording), and away from the corresponding feeling in your body. You can observe how it will do everything it can to keep you in this reactive narrative. This is the AH HA experience.

Your feeling presence is death to the mind, because going with and making peace with the feeling enables you to transmute both the judgment and the corresponding pattern of energy (the feeling) into the light of your true Self.

You have transformed what could have been a major upset into a joyful experience. You may end up laughing hilariously at the childish nature of your reactive mind. This is the HA HA experience.

Stopping and "counting to ten" has very little if any effectiveness when compared to re-creating the feeling. All that is, is the effort to try and control or change something, which absolutely *ensures* its persistence. You cannot create real resolution by "thinking it through" either. The best that can happen is that the unresolved feelings will go back into suppression; and you are still in the vicious circle.

When you transmute the energy on a feeling level, you are completing stacks of those endless-loop mind tapes. That leaves you with nothing, and that leaves you with space to create, to generate from your Beingness, from your connection to Source. You are also transmuting (disappearing) huge layers of stuck energy (the feeling component).

Re-creating your thoughts and feelings also puts you squarely in the present moment and gives you the opportunity to discover experientially that you are *not* your mind; you are not your thoughts and feelings and beliefs. You realize that all the things that truly matter—love, truth, beauty, joy, creative expression, and inner peace—all arise from beyond the mind.

You are waking up, you are stepping into your true nature as Infinite Consciousness, of no form awareness, having a human experience. Access to your true nature is available only in the present moment.

Now, what you create is a function of the total perfection of your Higher Self. Here you have no limitations, nothing is impossible, and what you create will not have its polar opposite as creation from the mind has. When you see it and feel it and know it with total certainty, it is yours.

The Breath will
take you deep
inside, to explore
a Self that's
vast and wide.

Breathe!

22
Breath Alchemy

As I have mentioned in previous chapters, there was a time in our distant past when we were in Unity. We lived the Oneness, seeing and living the One Spirit in everything. We had transpersonal memory, what was available to one was available to all.

When we fell thirteen thousand years ago, we move from Unity into separation; looking out at a world that wasn't us, as though we were all of a sudden separate and isolated entities, standing apart from the rest of the creation. Such is life when you become your polarized mind.

Your Higher Self is still within you, and it functions well beyond that of any capabilities that your mind might have. Reconnecting with this severed aspect of yourself gives you access to your unlimited potential; where you have an unlimited healing capability, along with reliable access to an infinite supply of inner peace, unconditional love, joy, creative expression, and wisdom. This is the energy of Divine Creator; it is the energy of Source.

When you know that everything is made of pure energy as I have shown you in previous chapters, you can truly transform your mind, emotions, heart, and body to bring about an abundance of health and happiness into your life. Breath Alchemy, based upon ancient wisdom, is a methodology that teaches you to work with this pure Source energy to access both power and wisdom.

Breath Alchemy is the art of expanding your sense of well-being, enjoyment, and benefit to include all of your life experience. It is an easy and enjoyable skill you can learn to do for yourself. It isn't religion or therapy. The forerunner to this process was known as Rebirthing, so named because the people who used it commonly remembered their own births.

There are three Universal Principles that are the foundation of Breath Alchemy. I introduced these three laws in the previous chapter, and here they are again:

The Perfection Principle: You are perfect, but there are barriers preventing you from experiencing and expressing your perfection.

The Resistance Principle: Resistance leads to persistence. If you try to resist or change something, it will become more solid. The only way to get rid of something is to just let it be. That doesn't mean to ignore it. Ignoring is actually a form of rejection or resistance. To ignore stress, anxiety, or anger is a very common way of trying to eliminate it. To let something be means to observe it, stay in touch with it, but make no effort to change it.

The Harmonizing Principle: The re-creation of an experience makes the experience disappear. Re-experiencing something to completion disappears it.

Breath Alchemy is the practical application of these three Universal Principles. It is a system that enables you to cause emotional resolution with regard to any stressful situation, working at the *feeling* level.

Content and Context

Let's take a moment now to look at the relationship between content and context. By content, I mean the thing itself or reality itself. The thoughts in your mind as well as the sensations in your body are all content.

Your relationship to what you are experiencing is called the context in which you hold the experience; it is the frame of reference in which you hold your experience.

Your mind does not relate to the content of your reality directly. Rather, it perceives reality through the lens or filter that is the context in which you are holding your experience. That means that your context determines how the content of your experience affects you. Knowing that that the circumstances in your life (the content) affect you only as a function of the context in which you perceive them, frees you from the trap of believing that it is your circumstances that determine your success and well-being in your life. Now the tail is no longer wagging the dog.

You are always holding the circumstances, the content in your life in a context, whether you know it or not, so why not choose your context consciously? Otherwise, your subconscious mind will choose it for you, which usually means taking what could have been a fresh new situation and contaminating it with stale old negativity.

A negative context is any context in which you compare the content to an imaginary standard and you tell yourself that reality should not be the way it is, that it should be the way you are imagining instead.

In other words, reality is the way it is whether you like it or not and you have chosen not to like it because it's not like the imaginary standard you have in your mind. That's a negative context.

The jargon term I use for it is you are "making it wrong." If you are comparing it to how you think it should be, rather than letting it be what is, then you are "making it wrong." "Make-wrong" is also a noun. The obvious implication of putting it that way is that the wrongness or unpleasantness of the experience is not inherent in the experience itself; it is a function of how you are making it. You are making it a negative experience.

A positive context is any other kind of context whatsoever, any context in which you are not comparing the content to an imaginary standard; this also includes a neutral context.

Whenever you are holding an experience in a negative context, you are not fully in touch with reality, because a substantial part of your mind is off in something that is totally imaginary and has nothing to do with reality. How it should be, how you wish it were, how good things used to be, what someone else has, etc. All that kind of stuff has nothing to do with the reality of your experience. That's all imaginary, so any time that you are comparing something to an imaginary standard, you are withdrawing from reality. You are not focused on the reality that is there; because a substantial part of you is focused on this totally imaginary thing, all your thoughts about how it should be.

Every Thought Has a Corresponding Feeling

Every time you make something or somebody wrong, you get an unpleasant feeling in your body; there is a negative thought, along with a corresponding unpleasant feeling in your body.

The reason that is true is because every thought has a corresponding pattern of energy, a feeling in the body.

For example: if you think about riding on a roller coaster, you get a certain pattern of energy in your body; if you think about your job, you get a different feeling in your body. If you think about what you want for dinner tonight, you get another kind of feeling in your body. If you think about voting, you get some other feeling in your body. If you think about your mother, you get some very complex feelings in your body.

Every thought has a corresponding feeling in the body, because *thoughts* and *feelings* are really the same thing perceived through two different internal senses.

Perceiving one thing through more than one sense is ordinary; if we were both in the same room for example, you could both see me and hear me. And you can hear yourself and see yourself and feel yourself. We are all accustomed to doing that.

We also have internal senses that correspond to our external senses. Everybody is internally visual to some extent and everybody hears words and sounds in their minds to some extent. Everybody also has an internal feeling sense.

It's not that the thought causes the feeling or vice versa, it's really that there is one thing perceived through more than one internal sense.

When you make something wrong conceptually (you can conceive of something better, so you make this thing wrong), the feeling component is made wrong at the same time. This is so because the thought and the feeling are the same thing; so the feeling that corresponds to the thought becomes unpleasant. Unpleasantness is just a feeling that you are making wrong. So that's why it is that any time you make something wrong, in every case, you get an unpleasant feeling.

These unpleasant feelings tend to become suppressed in the body; corresponding negative thoughts are suppressed in the subconscious part of the mind, where they are not experienced consciously until something in life reminds you of the make-wrong you've been trying to keep suppressed. Then the unpleasant feeling comes right back to your attention again.

Now let's revisit the age-old struggle between the Great White Brotherhood and the Great Dark Brotherhood. On the surface they are opposed to each other in every conceivable way; this is also true of the personal light-versus-dark battle inside you.

If you buy into the illusion of separation and "good versus evil," you are a victim in the situation. You will then be confronted by a very powerful Universal Law, one that works unerringly. It is the Resistance Principle; and it says resistance leads to persistence. This means that you will keep re-creating what (on a conscious level) you don't want; and it will only increase in intensity. You will be stuck in the mud; and the harder you spin your wheels, the deeper you go.

You will then tend to blame the unpleasantness you experience on whatever (or whomever) has just reactivated your suppressed negative emotions. You will then effort to control the situation and people in order to get the unpleasant emotions re-suppressed as soon as possible. Once you succeed at re-suppressing, you go back to being a ticking time bomb of emotions, hoping against hope that the world will cooperate with your self-delusion. Around and around this emotional merry-go-round turns.

It appears then, that the darkness is trying to keep you stuck by bringing the unpleasant feelings that you have been struggling valiantly to avoid. But that is only a function of your perception. It is actually just trying to get your attention so you can revisit these feelings and make peace with them so you can experience them consciously and integrate them.

All it takes to integrate the make-wrong or judgment is to go with the larger truth of Oneness, remembering that there is always only One Spirit moving through all of life, and that the light and dark forces within you are actually working together, acting as timing agents. This will align you with another powerful universal law known as the Harmonizing Principle, again one that works all the time: The re-creation of an experience makes it disappear, or in other words, re-experiencing something to completion "disappears" it.

And since Breath Alchemy is the practical application of the Harmonizing Principle; using it stops this vicious cycle. By voluntarily welcoming the suppressed feelings, the compulsive motivation to avoid them is eliminated.

That's valuable in understanding how this process works, because all it takes to integrate the make-wrong is to feel the feeling in detail. All you need is the physical feeling component of the experience, and to shift context with that, so that the feeling that had been unpleasant becomes a source of pleasure. As soon as you do that, you have integrated at every level and throughout all of your senses.

Because Breath Alchemy is both gentle and self-directed, it is easy to allow feelings to come up and integrate, regardless of how long they have been dreaded and avoided. It is called kinesthetic processing; it is done entirely at the feeling level.

Suppression

Suppression means choosing not to experience something that's there. Whenever you make something wrong, you get an unpleasant feeling in your body. It's normal for people to not want to feel bad, so they actually withdraw awareness from their body to avoid the unpleasant feelings that are there.

That's called *suppression*, and it happens because of two strong drives that everyone has:

1. Everyone has a drive to think that they are right about things. Your mind has to have that, because if you were continually questioning all of your assumptions about everything, you wouldn't be able to put two thoughts together very readily. You wouldn't be able to think about anything very effectively. So you start out assuming you are right about your basic assumptions. Yet there are disadvantages to that as I'm sure you have noticed.
2. The other strong drive is a drive to feel good. Once you have made something or someone wrong, and you tell yourself "That thing *really* is wrong," you *believe* your make-wrong; then how do you ever get to where you feel good again?

Most people stop thinking about it by distracting themselves (TV, food, drugs, complain, denial, gambling, etc. . . .)

This is called suppression, and it has three components:

1. Your mind will distract you from whatever has been bothering you.
2. Your body armors, it tenses up to give you less access to the subtleties of the suppressed feelings.
3. The breathing becomes inhibited, which for this process, is noteworthy.

What Breath Alchemy Does

Breath Alchemy brings breath and focused awareness together to empower you to heal yourself on all levels; mentally, emotionally, physically, and spiritually. Breath Alchemy enables you to produce superior outcomes by removing negative factors from your own subconscious. Yet Breath Alchemy works at the feeling level in the body, it is not a mental process. You notice your *feelings* about what's there, and you change your relationship to the *feeling*; and that's the most efficient method for changing your relationship to what the *feeling* is about.

Breath Alchemy clears up unresolved experiences from all periods of life, from before birth, throughout infancy and childhood, and right up to present time through a very specific result called integration. Everyone has a sense that certain things contribute to their well-being. But they also think that there are other things that also detract from their greater good. Breath Alchemy is about focusing on something that you think has been detracting from your well-being and by applying the process, you integrate it so that it is obvious to you that it does contribute to you.

Integration means that you shift to a positive context regarding something that you had been holding in a negative context.

Because of suppression, integration usually involves two components:

1. You bring the feeling out of suppression so that you are contemplating it consciously.
2. You shift from a negative to a positive context.

When you face the feeling squarely and accept it, you are integrating it. Integration is physical as well as mental. It makes you happier about being human. It also makes you markedly more effective at getting what you truly want from life because it eliminates internal conflicts.

Everyone has experienced integration many times (you get the monkey off your back, or you breathe the sigh of relief, etc. . . .) but for most people, the process of integration takes place rather haphazardly and they carry things that, after many years, they still don't feel good about. So, if you don't have a technique for creating integration, you're going to spend far more time suppressing than you are integrating. Therefore, it is greatly to your benefit to learn how to cause integration at will. Breath Alchemy is an *extremely* efficient method for creating integration.

This process then, is about harmonizing with the reality of your experience, *whatever* your experience is moment by moment.

How Breath Alchemy Works

Breath Alchemy uses a profound breathing method known as Circular Breathing to give you the experience of coming more fully into present time. The result of this can be a very useful magnification of emotional feelings, physical sensations, and spiritual feelings. When these feelings that arise from the subconscious are experienced thoroughly (which is not at all the same thing as intensely) and are embraced rather than pushed away, there is a profound sense of knowing oneself and loving oneself that makes all of life easier and more enjoyable. This learning to embrace is a process, however, and each person has a unique experience of learning to do it.

In your first session, you learn ways to achieve this new loving relationship within; yet, this is a genuine self-improvement technique. *You are the one* who decides how much loving nurture, benefit, or pleasure you get from any of your experiences. Full responsibility for developing a loving relationship with your feelings can only rest upon you.

The two main things the breathing does are to enhance your awareness of the energy streaming through your body, and to bring feelings out of suppression, making it easier for you to embrace them. It is *not* hyperventilation. You can easily learn to adjust your breathing to bring it to the perfect comfort level so you can integrate easily and on a subtle level.

All it takes then is to focus on the feeling as it is coming out of suppression by tuning it in, and then change the way you relate to it in order to accomplish integration. Integration does require having a thorough experience of the feeling. This is not the same thing as an intense experience—you can adjust your breathing to give you the level of intensity that suits you best.

There is no difference between changing the context in which you hold your feelings about something and changing the context in which you hold the thing itself. If you are truly at peace with another person, that is the same thing as being at peace with your feelings about that person. If you are out of harmony with that person in some way, you will have feelings that correspond to that. You will have feelings that you will also be out of harmony with. That is why you can integrate completely at the feeling or kinesthetic level.

Breath Alchemy is the Rolls Royce of all breathwork methods due to its efficiency, the ease with which it can be learned, and the pleasure with which it can be applied.

Breathwork has gone through major changes as it has evolved over the years. Originally it began as a very valuable but often painful and cathartic process involving hyperventilation (modern Circular Breathing *never* involves hyperventilation). The first book on Rebirthing, *Rebirthing in the New Age,* by Leonard Orr and Sondra Ray describes Rebirthing in its initial stages of development: dramatic, often intensely emotional or painful, but often ending with a "release" and transformation into a very blissful state.

Jim Leonard researched what happened in the moment of "release" wondering what exactly caused the bliss and why such a blissful result often happened only after discomfort. What he found was that at the moment of release people finally surrendered to feelings with which they had been fighting. He theorized that if possibly a person could surrender sooner, there would not be so much struggle and the technique could be more efficient.

His research showed him that this process of surrender consisted of five specific components: 1) Circular Breathing, 2) complete relaxation, 3) awareness of the details of emotions and sensations, 4) acceptance, love, or enjoyment of the details of the experience, and 5) knowing that integration can occur in the midst of what one is actually experiencing and doing in the moment. These components are known as The Five Step Harmonizing Method. The Breath Alchemy technique is using breathwork with these Five Steps.

I was originally trained in rebirthing; it was not however, producing the consistent results that I was looking for, so I went searching for answers. I found what I was looking for as a result of my decision to work closely with my friend, Jim Leonard, and his wife at the time, Anne Jill Leonard. Jim was the originator of the Five Steps (the Five Elements, as he called them). I first learned of his remarkable system a few years earlier; but it wasn't until late 1987 that I fully committed myself to mastering his technique. He was extremely generous in his willingness to share his time and expertise with me, and for that, I am eternally grateful.

This process is really about maximizing your enjoyment of the present moment. In the course of doing that, anything that is standing between you and enjoying each moment infinitely, will come to your attention. And you can apply the process to that.

How it is done is through the application of the skills that are known as the Five Step Harmonizing Method. The Five Steps then, are ways of relating to the present moment. The Five Steps are: 1) Circular Breathing, 2) Complete Relaxation, 3) Awareness in Detail. 4) Integration into Unity, and 5) Do Whatever You Do, Willingness Is Enough.

Every moment *feels* like something. Another way to put that is in every moment, there is some pattern of energy in your body that is *most prominent*, so the Five Steps are:

1. You breathe in the way that most enables you to have *rapport* with that feeling. Depending on what's coming up, you breathe a little

bit differently, or depending on your relationship to what's coming up.

2. You relax in the presence of the feeling, rather than taking action to make it go away.

3. You tune the feeling in. It is asking for your attention, so you give it your attention. Awareness in detail is about how to tune the feeling in so you can integrate it at a subtle level, without having to totally hype it up with the breathing. If you don't know about tuning in a feeling, you will have to wait until it becomes very intense before you will give it enough attention to integrate it. Integration does require having a through experience of the feeling. If you know how to tune it in, you can have a thorough experience of it at a very subtle level.

4. You find a way to accept the feeling as it is, rather than insisting that something else should be your present moment experience.

5. The Fifth Step is that whatever your response to the feeling, is all right. With your willingness, everything leads to integration any-way. So you don't have to do the first four Steps perfectly.

It is important to realize that the Five Steps are not steps that you do sequentially, instead you do them all simultaneously, because Breath Alchemy is just one process.

Master the Five Steps and You Master Breath Alchemy

Using the Five Steps helps people surrender faster to what they are experiencing in their breathwork sessions. The results are profound and immediate—without pain and drama, and with much more pleasure and more results in fewer sessions than traditional rebirthing and other breathwork techniques.

I teach people how to use the Five Steps in private sessions, before beginning the first session. This facilitates making the results rapid and very pleasurable and reproducible. In these experiences, one is authentically connected to their sense of wholeness and unification within themselves and with all of life.

Because the Breath Alchemy can be used to clear negativity of any type or origin, it is quite possible to learn to guide yourself while engaging in other activities, at work for instance. This means that in day-to-day life, emotions can be used to one's benefit (and indeed converted to pleasure) *as they arise for whatever reason.* In general, my goal is to develop that level of autonomy in each person as quickly as possible. An intermediate step is the ability to complete a session without my assistance.

It is important to realize that the Five Steps are skills; they are not a model of what happens naturally in the Breath Alchemy Technique. All of the Five Steps are skills that you have some degree of mastery over already, just because of your life experiences. At the same time, they are skills that you can keep getting better at forever.

Breathwork Without the
Five Step Harmonizing Method

Learning Breath Alchemy makes an extreme positive difference. Although breathwork without the Five Steps does work, it is less efficient and can be problematic. Without the aid of the Five Step Harmonizing Method, developing autonomy skills can take much longer.

Without knowing how to adjust the breathing, for instance, the experience can be sometimes too subtle and sometimes too intense to create optimum results. Or, if you don't know how to relax or how to broaden your positive contexts at will, you're in for needless struggle and wasted time. If you don't know how to "tune in" to the physical and/or emotional feelings, you'll either have to wait until they become very intense or else waste time "waiting for something to happen."

Breathwork is not a process of drama and confrontation unless the facilitator thinks it is. Focusing on catharsis or confrontation during a session can even reinforce the negative thought that unpleasantness is necessary in order to make improvements. Inevitably, these dramatized processes reinforce the idea that there is something wrong with you that must be gotten rid of. Breathwork can be extremely pleasurable throughout and indeed is at its *greatest effect* when it is at its most pleasurable, as the Breath Alchemy Technique will teach you.

The Five Steps eliminates the all the guess work entirely. You can easily learn to apply the Five Steps in one session with my guidance. With a little more preparation time before launching into the session itself, it will make every session you have that much more efficient and comfortable, and is well worth it

The main thing to keep in mind—to fully appreciate about Breath Alchemy is that it connects the brain with the heart where integration into Unity happens. The Heart Mind (the Higher Self) knows what to do, and the mind/body/emotions/energy are harmonized.

This is Key.

In addition to being the direct connection to your higher guidance, Breath Alchemy has many additional advantages: 1) Your breath is readily accessible to you. 2) When you have the ability to breathe prana, it becomes easy to integrate any physical symptoms or negative thoughts. 3) It is safe; breathing and relaxing can never hurt you. 4) It is a kinesthetic process (which means that the focus is on integrating into your sense of well-being *all* sensations/feelings in your body). Kinesthetic processing is faster and more direct. A single feeling can be worth thousands of words and concepts that would take hours to talk about. You can rationalize many situations in your mind, and your mind can keep you in denial for years. Your body doesn't lie. Feelings are much harder to deny, and feelings let you know the results of your unconscious and conscious thoughts immediately. 5) Breath Alchemy can be used to enhance any other self-improvement technique, therapy, or treatment.

Once you learn how to use my Breath Alchemy technique, you can use it by yourself anytime.

Twenty Circular Breaths

The foundation of Breath Alchemy is a simple exercise called twenty circular breaths. You can do this exercise throughout the day, whenever you feel the need.

1. Take four short breaths.
2. Then take one long breath.
3. Pull the breaths in and out either through your nose or through your mouth.
4. Do four sets of the five breaths, that is, four sets of four short breaths followed by one long breath without stopping, for a total of twenty breaths.

Circular breathing means any kind of breathing that fits the following three criteria:

1. It means that the inhale and the exhale are connected together without any pauses. As soon as you complete the inhale, you abruptly relax, and let the exhale just flow out. When the exhale has gone as far as it will go, you immediately begin the next inhale. So there are no pauses.
2. The second criterion is that the exhale is completely relaxed, and not controlled at all. You don't use any muscles to make the exhale take longer, or any muscles to try to squeeze all the air out. As soon as the exhale has gone as far as it will go with relaxation alone, you immediately begin the next inhale.
3. The third criterion is that either you breathe in and out through the nose, or you breathe in and out through the mouth. Either one

of those works, but if you switch back and forth—if you inhale through the nose and exhale through the mouth or vice versa, it weakens the effect of the circular breathing quite a bit.

Now, what is the effect of the circular breathing? The simplest way to explain it is that it gives you complete circuits of energy in your body in a way that enhances your awareness of the life-force energy (prana) in your body.

Use the short breaths to emphasize the connecting and merging of the inhale and the exhale into unbroken circles.

Use the long breath to fill your lungs as completely as you comfortably can on the inhale, and to let go completely on the exhale.

Breathe at a speed that feels natural for you. It is important that the breathing be free and natural and rhythmical, rather than forced or controlled. This is what enables you to breathe prana as well as air.

Since most of us have developed bad breathing habits you might experience some physical sensations, such as lightheadedness or tingling sensations in your hands or elsewhere. If you will tune these feelings in and make peace with them, you will notice that the sensations may change and become less overwhelming, and more generative of healing. This indicates that you are learning about breathing consciously and are getting direct benefits in your body.

I must emphasize that this is only a taste of the Breath Alchemy Technique. An actual guided session takes you light years beyond anything this beginning exercise can give you. If you wish to accelerate the process, contact me and schedule a series of one- to two-hour guided Breath Alchemy sessions. Sessions are conducted on Skype; I can be reached at BobFrissell.com.

The majority of people take from two to twelve sessions to feel safe, trust the process, and develop their skills of breathing, relaxing, focusing, and finding positive contexts. I have experienced the energy sensations many times that come up during a Breath Alchemy session and

I have the knowledge of how to adjust the breath and how to create a setting of total safety with regard to those sensations. I am a breathing guide who allows and trusts *your own* life energy to do the session. In truth, you control the session. I create an environment of safety for you to experience your own divine energy and bliss. I can telepathically pick up what the energy is doing in your body and channel energy to make your session more powerful. However, my main goal is to teach you how to give yourself unassisted sessions.

Once you learn Breath Alchemy, you can use it anywhere while engaging in any activity. You can use it at work in order to integrate any emotions that come up into your sense of well-being so that you can have all of your feelings and all of the situations that arise, contribute to your creativity, to your effectiveness, and indeed, to your happiness. You can use it while driving, you can use it while you are talking on the telephone, you can use it while you are watching television; you can use it anytime, anywhere.

Frequently Asked Questions

How long does a session last?

There is no prescribed length. Usually, coached sessions last from about sixty to seventy-five minutes. It is important that you go until you have completed an energy cycle and you feel great, regardless of the time it takes. Almost all sessions are somewhere between one and two hours. Sessions are cumulative; they build upon each other. After your skill level increases the sessions usually become shorter.

What kind of changes can I expect?

Circular Breathing creates a new muscle memory, and the Five Step Harmonizing Method creates a new emotional response. These two things together will give you a new paradigm for dealing with stress in your life as it comes up in the moment. You can therefore expect greater relaxation in formerly stress producing situations, freer flowing creativity, willingness to take action to resolve situations you have felt stuck in—and to do so with clarity and confidence, greater feeling of presence and connection with friends and loved ones.

How long should I keep doing Breath Alchemy?

This is not a "quick fix" solution. Breath Alchemy is a process for harmonizing the emotional and spiritual energies of the body. It can be applied in dedicated sessions and also in day-to-day life, during both pleasant and unpleasant experiences. When experiencing pleasurable situations and feelings, it allows you to be more present and receptive. When you are dealing with something difficult, Breath Alchemy relaxes you into a more creative perspective and frees you from whatever negativity was keeping you stuck. It is a technique that you will have for as long as you walk this earth that you can use to increase your enjoyment of life and your effectiveness.

Can I do Breath Alchemy on my own?

Yes. I can teach almost everyone how to do self-Breath Alchemy sessions in five to twelve coached sessions. Each assisted session has a two-fold purpose: 1. To make sure you have completed the energy cycle so you feel great, and 2. Each session is also a lesson in how to give yourself solo sessions. I will also teach you the 24/7 aspect of Breath Alchemy by giving you some simple daily techniques for shifting any experience from a negative context to a positive context.

How do I know if I integrated something?

You can *feel* the energy transmuting in your body. You will experience integration differently at different times: sometimes you will feel more energized and alive and present and connected, sometimes you will have a profound sense of relaxation, and sometimes you can feel the emotional energy you had been experiencing as unpleasant suddenly become a big source of pleasure. Other times something will cease to be an issue so completely that you'll forget it ever was an issue.

Will I know what I've integrated?

Sometimes you will have a clear understanding of what has been resolved; other times it will be a purely feeling level integration. You do not need to have a cognitive understanding of what it is, because Breath Alchemy works entirely at the feeling level; it is not talk-therapy. It is *extremely* efficient, because when you have integrated at the feeling level you have integrated on all levels and throughout all of your senses. For example, if you are truly at peace with your feelings toward your mother, that is the same thing exactly as being at peace with your mother herself. If you are out of harmony with her in any way, you will have corresponding unharmonious feelings toward her.

Do you have a different experience every time?

Yes, things that get integrated tend to not come up in sessions anymore, unless you make them wrong again, which you can. But because integration works at both the feeling level, and the conceptual level, you are unlikely to put the make wrongs back again. So yes, it is different every time.

Is integration permanent?

Yes. Making the sensations in your body wrong and trying to keep them suppressed requires effort. Your mind and body seek peace. Every integration is a quantum leap in receptivity to the benefits, pleasures, and learning available from something in your life. Each integration invites your body to present you with the next feeling. As you develop skill with the technique you can cause many integrations very quickly, and in each session, you experience a healing adventure into the depths of yourself.

How will I know when my session is complete?

Whatever has been interfering with your ability to fully enjoy your aliveness comes to your attention sensation-by-sensation, and is transmuted; it is integrated. Upon reaching the point of completion for the session, you will feel calm, peaceful, and alert. If you are feeling great, energized, and relaxed, then your session is complete.

How do I contact you for more information and/or to schedule a session with you?

All Breath Alchemy sessions are conducted on Skype. Prospective clients begin with a free thirty-minute Discovery Call. The purpose of the call is to create a vision for your life that will inspire you to take the first step to move ahead, identify the true source of your stuckness so you can shift from frustration to focused action, and get crystal clear on your next step to moving ahead with clarity, precision, and confidence.

After we have talked about all of the above, if we both feel that it makes sense for us to work together, I will share with you what that would look like. The best way to reach me is to send me an email.

A Note from Drunvalo

In the ancient schools, such as Egypt's, the female or right-brained aspect of the Mystery School (the Left Eye of Horus) always came first. The student began working on his or her emotional healing, and after the emotional healing took place, then the left-brained aspect was taught (the Right Eye of Horus).

Here in the United States, and in other left-brained countries, we introduce the left-brained studies first without the emotional healing because these countries are having a difficult time understanding the female pathway. In many cases, they have flatly rejected this simple way. Therefore, we have introduced this male pathway first (just to get their attention). But now that we have your attention and you are beginning to study this pathway, I find it necessary to say that you must know (or at least at some point on this path begin to study) the female way.

Emotional healing is essential if you really wish to find enlightenment in this world. There is NO way around this. Once you begin to find out about the higher worlds, you yourself will stop your own growth past a certain point until this emotional healing has taken place. I am sorry, but that is the way it is. You can't do the Merkaba meditation or any other kind of meditation to any real degree of success if your emotional body is out of balance.

My suggestion is that you trust yourself and open to the possibility of someone coming into your life who can help you with your emotional imbalances (even if you are not aware of them). It almost always requires help from the outside. We usually cannot see our own problems, and so this is one area of human experience where outside help is just about the only way.

Only when a person is in a relatively healthy emotional balance can they successfully function through the Merkaba.[1]

23
Tapping into the Creative Flow

In the introduction I told you of a decision I had made—that the only way I would ever write a book was if a publisher came to me and asked me to do so. And though I thought the chances of that happening were somewhere between slim and none, that of course is exactly what happened.

My first thought upon receiving this fateful phone call from Richard was "No, I'm not an author." I then remembered my decision of four years earlier, and I realized that there were higher forces at work here, and that meant that "no" was not an option.

So now I was committed to writing a book.

Now let me ask you, "Have you ever found yourself in the face of a task that seemed so herculean that it just seemed impossible? That no matter how you tried to tweak it, it just seemed well beyond your reach?"

Well, that's exactly how I felt; I didn't know how to write an article, let alone a book. I didn't even know how to type! So, what to do, and where to begin?

I decided to start with what I knew best, and since I was a teacher of breathwork (this was well before I discovered the more appropriate name of Breath Alchemy), I began with that. Start with what you know best, sounds like a good idea, right?

That is until I tried to put down what I knew on paper. So even in the area of my greatest expertise, my attempts at completing a chapter went from bad to worse to "This is impossible, I can't possibly do this!"

Not the best way to get started on a new book I thought.

"This is impossible, I can't possibly do this" was my consistent over-riding thought; the voice was loud and obnoxious and I woke up to it every single morning.

So, what did I do? Well, I decided I'd better practice what I preach.

One of the greatest motivating factors for me over the years has come from my relationship with my clients. They would tell me of the successes they were having, and that in turn inspired me—whenever they seemed to be getting ahead of me—to go for that level of accomplishment for myself.

"There are no accidents, and the universe will give you everything you need as long as you are committed enough to send a clear message of what it is that you wish to create."

That came from me; it was the message I gave to my clients to help them in achieving distant and far-reaching goals.

I knew it was doable for them because every session resulted in integration of suppressed material that had been limiting them and holding them back. So, by removing the barriers, they were now free to move ahead. That is really the same as saying that every completed session resulted in a creative breakthrough for them; it allowed them to shift from "what's missing" to creatively utilizing whatever resources that were available to them.

I flashed back to four years earlier when a new client came to me. She told me how she felt stymied; she was at a tipping point and she didn't know which way to turn. She said she felt damned if she did and damned if she didn't. She told me of her yearning to make the right decision, one that is admirable that has integrity, and is in the best interest of all concerned. But she just didn't know what to do.

What she was referring to was an impending divorce and all its ramifications—what about her three-year-old son, what about her career, where would she live, and so on.

I wasn't there to advise her on how to proceed; my job was to show her how to discover and master her innate ability to resolve suppressed emotions that had been limiting her and keeping her stuck. And the answers would then come from within; they would come as intuitive insights from her newfound connection with her truest self—her Higher Self.

So, what happened?

It's funny how there are no coincidences, that life will give you everything you need if you're open enough to see it and to receive it.

As it turns out, she was an author, the same person who upon showing me the detailed notes she was taking of our work together, suggested we co-author a book. Even though I declined her offer (truth be known, my fear of writing took over, and I didn't think I really had anything to say), I was impressed with what I was reading—so much so that I still have my copy to this day.

Her relationship with her husband began to turn around after I told her whatever emotions she had been suppressing were going to get activated by her partner, and in fact, she could count on that. Then she mentioned that he seemed to be constantly doing things that agitated and upset her; he made her angry is how she put it. So I used anger as an example. I told her if the only way she knew how to relate to anger is to make it wrong, then if her husband does something that activates her anger, she is not going to be very loving of him right then. She readily agreed with that! I said she will be blaming him for triggering this emotion that she is resisting; she will be trying to control his behavior to not activate this emotion anymore.

Then I said, "If you know how to integrate your anger, you can love him even when your anger is activated. If you allow all of your emotions to cycle through and be okay with you, and you are able to include all of your feelings in your experience of love, then you can genuinely be unconditionally loving. Otherwise, you can only love with the condition that he not set off the emotions you are choosing to resist."

She also had outstanding issues with fear, guilt, and shame, all of which got resolved in the most efficient way possible—at the feeling level.

Let me state for the sake of clarity that because breathwork (or Breath Alchemy as I now call it) works at the energetic level, it is effective for anything you can feel. That includes a huge list of possibilities, ranging from headaches to stress to fatigue, with many stops in between. Any sensation that has been bothering you can readily be transformed or integrated using the Five-Step Harmonizing Method as described in the previous chapter.

Breath Alchemy is also an extremely efficient method for integrating emotions. It does so in the most direct way possible; you simply focus on the emotion you're feeling and find a positive way of relating to it. Each emotion is made of energy, and the energy it contains is held away, as a separate entity, until you integrate it. Once you integrate the emotion, its energy merges back into your overall aliveness. In the same moment, you gain the lesson the emotion has for you.

Fear for example integrates into excitement and alertness. Anger integrates into intention or determination. Integrating guilt enables you to forgive yourself and others so you can move on. Shame is much more complex because unlike guilt, which is you made a mistake, the message behind shame is you *are* a mistake! Furthermore, since we live in a shame-based society, all your emotions will be bound with shame. Therefore, there will be an integration of shame every time you integrate an emotion. Integrating shame-bound emotions is one of the most important keys in reclaiming your childlike innocence, which is essential in the process of reconnecting with your Higher Self and your natural divinity.

Not only did her failing marriage seem to magically resolve itself, but she also said she was feeling more creative in her writings than ever before. What she was saying was that the stress and anxiety she was feeling in her personal life was also creeping into her professional life and keeping her stuck in what seemed like a constant brain-cramp mode with her writing too.

Now she said the chapters to her current book just seemed to write themselves. And that newfound creative flow was very evident in the writings she showed me.

And these creative breakthroughs came after only a few weekly visits. Then her husband came to me for a series of sessions. He told me that with their newfound ability to integrate their emotions rather than act them out, it had become clear to both of them that once they got beneath all the resentment and all the hostility, and all the blame and the guilt, and the shame—that beneath all that stuff, that they did indeed love each other. And they were both grateful, he said, to be given this second chance.

So just like in all good tales, this one too has a happy ending. To this day, it lives as one of my favorite success stories. It was a real game-changer for all three of us!

Now here I was four years later feeling the same or perhaps an even deeper level of stuckness than she had felt when she first came to me. And I'm the expert, right? Could the master humble himself enough to recognize that his students might well be his greatest teachers?

If it worked for her, could it possibly work for me???

That's when I made the decision to do whatever it takes, and if that meant giving myself a breathwork session every morning, well so be it.

So I gave myself a daily session in direct response to my kicking screaming mind that was telling me that I couldn't do this. It told me I'm not an author, I don't have the necessary expertise, and so on, ad nauseam. Every completed session resulted in an integration of a deeper layer of my fears that were keeping me stuck; it also resulted in the appearance of a peaceful, much softer voice that was saying to me, "You can do this, the way will reveal itself."

The way did indeed begin to reveal itself and I quickly realized the need to come from this balanced integrated place. I remembered when I was just casually and enthusiastically sharing with Richard, how the words just effortlessly flowed out of my mouth in a way that certainly seemed to captivate him and hold his attention. So I got the idea of talking informally into a cassette recorder and then transferring that to paper. And though it was time consuming, it worked. Then as I began to let go of any need to "get it right," the words just began to flow through me. I was now beginning to grasp the possibility that I could indeed do this!

I also remembered the famous quote of the Chinese philosopher Lao Tzu: "The journey of a thousand miles begins with one step." I realized that every single action was a contribution to the completed project, so I began to proceed, one step at a time, and one day at a time. And the rest as they say, is history.

This is how I came to fully realize that every integration—the intended result of every breathwork session—also results in a creative breakthrough. Now it was time to reverse engineer my discovery.

Though it was far from obvious to me at the time, it finally dawned on me that "This is impossible, I can't possibly do this," was a pretty clear example of a judgment; and each complete and integrated cycle in my daily sessions resulted naturally in a shift to a positive context. Or in other words, I moved from being stuck in what's not working or what's missing, to being able to see and to utilize whatever resources that were at my disposal. That in turn gave me access to my internal creative flow.

From this I realized that even though writing a book was indeed a steep hill to climb, I had been turning a challenging but achievable project into an impossible task as a result of my inappropriate choice of contexts. I had taken something that could be seen as an exciting adventure and turned it into a monstrous undertaking that seemed well beyond my reach.

The first step, I then realized, is to make it okay that you have a situation at hand that needs to be resolved. Nothing as I discovered, inhibits the creative flow more than being stuck in the frustration and the impossibility of it all. Even if a solution did appear, it would most likely be passed over due to the judgmental context in which the situation was held.

Let me illustrate this by way of example; I'll cite the classic case of being in a hurry to get out of the house, but you can't leave because you can't find your keys. Even though they're obviously not where they "should" be, as the sense of frustration builds, you continue searching for them in what has been their usual place. It's not until you somehow find a way to accept the situation as it is, that your negative feelings will subside enough to allow creativity to return that will enable you to consider other places where they might be. Then as is often the case, they just "magically" appear.

It had also become totally obvious to me that many uncomfortable emotions, namely fear, frustration, hopelessness, and helplessness would become activated as a result of undertaking this endeavor. Any of these feelings can be powerful enough to stop you in your tracks, so knowing how to integrate them was critically important. I developed an enormous sense of gratitude and appreciation for my ability to integrate them in my sessions, and also on the spot.

My fears integrated into excitement and enthusiasm, while the frustration integrated into an increased determination to complete the job at hand. Hopelessness and helplessness transformed into powerful intention. This emotional resolution also left me in the present moment where I was able to proceed one step at a time.

And speaking of excitement, even though it happened completely by word of mouth, my book became an overnight sensation. This began to fully dawn on me when one day I casually strolled into a tiny bookstore in Marin County. It was a place called Paper Ships. To my surprise, I saw my book prominently displayed, so I introduced myself to David the owner, and I listened intently as he told me of the enormous and growing popularity of my book. It was, he told me, their number one seller. Then I went to a much larger store and was completely blown away as I looked at the largest display in the store. What was the featured book? It was none other than *Nothing In This Book Is True, But It's Exactly How Things Are.*

24

Your Body Is Light and You Are Immortal

Let's begin by taking a closer look at the title of this chapter. Even though it certainly appears to be, your body like everything else, is not physical; it is energetic, a holographic illusory physical form, which in its base-state is waveform information. This base foundation vibrating energy, containing massive amounts of information, is decoded into the apparent physicality of our bodies through the decoding system in our head, our heart, and our genetic structure.

This illusion of apparent physicality and solidity is furthered by the density of our third dimensional reality, where things do indeed appear to be solid. However, if a fourth dimensional being were to suddenly appear in your room, this truth would be readily apparent—he or she would appear as a being of light, capable of walking right through walls.

And then there's the issue of immortality—and yes, I've softened you up with all the talk of Thoth and the ascended masters—and remember, these guys are still alive today. Let's take a moment to recall that there are two alternatives to death as we know it. But first, let's consider what happens when we die in the manner in which we have become accustomed. The lower overtones of the fourth dimension are rather disharmonic, and that's where we go when we die—specifically to the third overtone. But we do it unconsciously, so when we get there, we have no memory of our lifetime here. For the past thirteen thousand years we have been cycling back via reincarnation, and we have been doing that unconsciously too. So, when we come back here through birth, we have no memory of where we were or of our previous lifetimes.

The two alternatives to physical death are both eyes wide-open options. The first, resurrection, is consciously moving from one world to another by dying and then reforming your light body on the other side. With ascension, you don't die at all; you consciously move from one world to another, taking your body with you. It is a very responsible way of leaving.

There are two main advantages with these alternatives. First, you don't stop at the third overtone. You continue until you reach the higher, harmonic overtones of the fourth dimension—either the tenth, eleventh, or twelfth. These are usually referred to as the Christ-consciousness overtones. Second, you are able to keep your memory intact; in fact, you will never again have a break in memory.

But with that being said, the topic of immortality is still too far over the line for most people. It's okay to talk about longevity—say to live healthfully to 120, but immortality is a dirty word. Growth hormone, meditating, or eating organic food—yes, but not immortality! Yet in spite of this modern-day taboo, a fundamental fact remains, and that is we were initially created to live forever—in our bodies!

In my attempt to get to the bottom of this, I thought the Bible would be a good place to look. I am not a student of the Bible, but I discovered in our many conversations with a friend and client by the name of Wilbur Albrecht—who was a religion major turned doctor—that he indeed is! So, I had a lengthy talk with him for the purpose of gathering information for this chapter; and we came up with the following.

Adam and Eve were immortal and in a state of innocence which they lost by eating from the tree of the knowledge of good and evil (the knowledge of the mind, the knowledge of polarity consciousness). Adam and Eve were created and designed to live forever, but they became mentally polarized and no longer heart-centered. It was their inability to remain attuned to Source that caused them to fall into mortality.

You can say this whole immortality thing is very esoteric and very difficult to get people to look at, however, the Bible is the bestselling book in the world; we are still a Christian based society—we still use the Bible in court. And the story of how immortality was lost is all there, it's right in our face. You can actually point to the Bible and say the model is that allegory.

The patriarchs in the Old Testament continued to fall in terms of their longevity, so new covenants were brought forth. Why? Because mankind was no longer able to stay attuned to Source and was out of alignment in that regard. Covenant, according to Wilbur, is a theological term for scholars when they look at the Bible. Covenant means that Source (the higher-ups, whether they were star beings or ascended masters or whatever they really were back then) was bringing forth a new way for humankind to have a new relationship with the Divine so they could be attuned to Source again and not be in sin. Sin as you may recall, is an archery term meaning missing the mark—you're not aligned anymore.

So, Abraham had a new covenant; and then Noah with the flood was the same deal. Humankind wasn't able to come into alignment with Source; there was a continual falling of consciousness, so there was a new dispensation, and that's what we saw with Noah. And then there was Moses, who lived to one hundred and twenty. He brought the Ten Commandments.

Jesus demonstrated transfiguration; he turned brilliant white, he became full of light. He demonstrated resurrection and ascension.

Pentecost came, which is the descent of the Holy Spirit, which is our capacity to communicate with our Higher Self. This is a new dispensation, directly communicating with the Higher Self or Source. Then Jesus ascended, and that's what really bummed out the disciples—he took off, he literally ascended.

Yet today with all the people doing yoga, meditation, breathwork, energy work—if you bring up the topic if immortality, you're going to the far limb of the tree. People can't seem to even consider it, let alone accept it. However it's always been right in front of our faces and we can't see it. But just look at the Bible—it's all there—Adam and Eve, made to live forever, the continual fall in consciousness, the continual requirement for new dispensations, the continual decline in longevity. So we can't see the forest for the trees; it's right there.

Wilbur then went on to say that it was only through working with me and getting my approach to higher consciousness and immortality, that he was able to see it plain as day, that the Bible is really the story of this immortal couple who were created by Divine consciousness, and there's a continual decline in consciousness. As a result, war and death became popular. And it culminates with the man who demonstrates transfiguration, resurrection, and ascension. Jesus was a Master from a much higher level, who came to re-put the truth back into the Akashic records. And still we missed the mark; we used it to create even more ego and separation.

So now I'll tell you what I learned that in Wilbur's words, assisted in propelling him over the finish line so he could plainly see.

For me, it all began with my burning desire to heal myself from a serious crippling and disabling lower back injury; which at its worst, left me turned and twisted and riveted in place like a pretzel. And yes, it was painful! One day in 1983, I got a call from a friend in Santa Cruz, California, who told me that she had a group together, and she wanted me to come down and give a weekend workshop. I told her of my condition—of how it would take me forty-five minutes or so to make any kind of movement—to get out of bed, to stand, to do anything and that driving was impossible. She said she understood and then suggested that since I was feeling miserable no matter what I did, why not come down and get paid for it? Then she sealed the deal by offering me a two-way ride.

I got plenty of help from chiropractors, but when my condition had worsened to this extreme, it took two weeks or more of daily visits to get me back into some semblance of normalcy. And why not doctors, you might be wondering—I had heard enough before and after horror stories of back surgery and its resulting complications, leaving the patient in even worse shape. So, the law of diminishing returns had entered into my experience with chiropractors; and since doctors were not an option, I was left with only one alternative—and that was to find a way to heal myself.

I first heard the idea of physical immortality in March 1979 when I was introduced to rebirthing (a breathwork technique developed by Leonard Orr, and a forerunner to my Breath Alchemy Technique). While it initially did not compute, I was well aware of the fact that I needed every bit of power I could muster, lest I fall into victim mode. So I quickly saw it as a most useful context in which to hold my injured back. I realized that as a container, it was broad enough and totally inclusive of our true nature; and that is of spiritual beings having a human experience who were initially created to live forever. By adopting this context, I was now able to reduce my injury, which had been out of my control, to mere content; and as such I now had access to the innate power within me to heal myself. Even though it took me twelve more years to internalize enough to be able to completely heal myself, that was a major turning point; it put and kept me in my power, by supplying me with a context that was inclusive of everything that came up for me. My thanks go to Leonard for introducing me to the idea.

Since Wilbur had been forced into early retirement as a burn-out victim with a multitude of health issues, including nearly losing his eyesight, I quickly and continually impressed upon him the idea of adopting an all-inclusive context that would put him in control of his healing process. The results were nothing short of amazing. I am including a summary of his accounting of the many benefits he received in one years' worth of work with me in the next chapter.

According to Leonard Orr, there are five immortals of the Bible. They in addition to Jesus, are Enoch and Elijah—both of whom ascended after nine hundred plus years, Melchizedek who had neither birth nor death—meaning he materialized an adult body, and did so at the time of Abraham. And then there is Moses who at the age of one hundred and twenty, if Leonard is correct, either resurrected or ascended.

In May 1980 I took a seminar from a couple named Sue and Flori Riggs; then I spent a day with them and received a rebirthing session. In this session, I felt life energy returning to my low back, if only for a brief but shining moment. I had completely it seemed, withdrawn all feeling awareness of that area to the point to where it felt almost as if my body was severed into two pieces. I could feel the areas above and below the injury, but not the area itself. That session convinced me that breathwork held the key to my healing; however, it took seven additional years of struggle to realize that a more developed form of breathwork was needed. That led me to Jim Leonard, and the rest as they say, is history.

I also learned from Flori, that his grandfather, Sam Riggs—a Kentucky farmer who was born in 1850—had in his possession some rather extraordinary talents, which included the ability to materialize wheat and other foods for his family. He also, according to Flori, was able to dematerialize his body. Flori told how Sam had made several pilgrimages to Death Valley, California, and did so for the final time in 1937. Before departing he told Flori that if he ever needed him, he would return. He then turned away, walked between two trees and disappeared.

There is a classic set of books called *Life and Teaching of the Masters of the Far East* by Baird T. Spalding. It tells the story of a research party of a team of eleven scientists, who were trained to accept nothing unless it was fully verified, and who never took anything for granted. The books give the details of their three-and-one-half-year visit to the Far East, beginning in 1894. During their stay, they were contacted by the immortal masters of the Himalayas, who aided them in their translation of ancient records. This was of great assistance to the team in their research work. Though quite understandably skeptical at first, they came away thoroughly convinced, so much so that three members of the team returned; they were determined to stay until they were able to perform the works of and live the life of the masters.

These masters also permitted them to intimately enter into their lives; and in their many travels and experiences, regularly materialized all supplies out of thin air, demonstrated their ability to walk on water, walk through fire, and dematerialize and rematerialize their bodies— thus enabling them to travel distances of many miles in an instant. Many of these individuals, all of whom had the power of ascension, were in excess of five hundred years of age. And they all insisted just as Jesus did, that we all have the same powers and abilities contained within us.

Meanwhile back at the ranch, there is another important factor that I would like to include; and that is the idea of the unconscious death urge. The death urge was brought forth not only by Leonard Orr, but also by Sigmund Freud, who clearly talked about the death wish; he said the pain of living is so intense that we have all these defense mechanisms to guard ourselves from all the suffering. So difficult to grasp ideas can often be found in mainstream circles.

Anyway, the unconscious death urge is a real psychic entity that literally can be isolated in your own mind and destroyed. It is composed of anti-life thoughts and beliefs, and it is held in place by the belief that death is inevitable and out of your control. Its purpose is to kill you, and that is exactly what it will do, unless you kill it first.

If your thoughts, feelings, and actions unerringly create your reality, then the unquestioned, unexamined death urge that you inherited as a multigenerational pattern from your parents and from the culture will become an unwitting context for your life, and it will produce its intended result. There is tremendous power in simply questioning the inevitability of death. Consider that the source of victim consciousness might just be the belief that you have no say in the health, well-being, and destiny of your body. The ultimate victim is someone who believes it is someone or something "out there" that has control over him or her.

If you want to take control of your life, you must include the possibility that it is you who creates your safety and well-being and no one else. It is you who creates your own health, aliveness, illness, injury, and accidents as well as the death of your physical body.

So the contradiction is that since ascension is a function of full aliveness, how can you be fully alive if you haven't unraveled your death urge? If you are unaware of it, it's unconsciously producing results.

Furthermore, the more enlightened you become, the more activated the death urge becomes. Anything you are still subconsciously holding onto that is less pure than your highest thoughts is fair game to come to your attention. And since you are becoming more conscious, your thoughts are becoming more powerful and they will manifest more quickly. If you have never questioned death, you are unwittingly at its effect. You will probably rationalize too and conclude that since you will be meeting your maker, it must be for the highest good. If you really want to meet your maker, go for full aliveness so you can ascend!

Immortality is not about living forever in the same body—that is ultimately a trap. There is always a higher reality to evolve into.

Immortality is about keeping your body in a rejuvenated state for as long as you want to, then consciously leaving when you want to, and retaining your memory as you ascend through the heights. It is about remembering who you are, always. The ascension process begins when you move from your mind back to your heart.

25
Healing

I was motivated to learn the principles I am about to share with you out of necessity. As I shared in the last chapter, I was suffering from a chronic low back injury that had taken a number of turns for the worse, leaving me quite helpless for much of the time. I was committed to finding a way to getting my life back, no matter what and no matter how long it took. I clearly recall the many times I would stand in front of a full-length mirror and affirm with total conviction, "I want my body back!!!!!" I would then see myself exactly as I wanted to be, hiking running, doing whatever I wanted—and always in perfect health. You might think that with such a burning desire that a solution would be forthcoming, but not so. I was struggling to stay above water in a sea of confusion for many years, but I never gave up.

I didn't realize it at the time (try to explain water to a fish), but my condition was entirely stress induced, and underlying that was a sizeable reservoir of unresolved emotional trauma dating back to childhood. And it was held in place by a constant fear that it would only worsen, and for years—it did. Stress and anxiety were my constant companions, leaving me struggling to swim upstream in a deep sea of pea soup.

I had an experience in August of 1980 that unbeknown to me at the time, held the keys to my healing. This experience was so remarkable that to this day, I sometimes shake my head and wonder if it really happened.

Freddie, an adorable orange tabby, was found by me in a garbage can choking on an overly tightened flea collar. I rescued him and instantly adopted him; he quickly became my best friend in many ways. He was clearly grateful for having found a home where he was finally treated with the love and respect that he deserved. He was playful and affectionate—he was always good for a hearty "meow "whenever I put my hand on the top of his head to pet him.

One day Freddie became quite paralyzed; he had absolutely no use of his rear legs. He was able to move around by inching himself forward with his front paws, and he wasn't in any apparent pain or discomfort. He was just being himself as though there was no problem. After taking him to the veterinarian who offered no solution, and then just hanging out with him for a few days, I felt quite helpless. I wanted to help him, but what could I do? Well one early evening I was sitting with him on my deck and all of a sudden, I had a totally spontaneous experience. I had a vision that quickly widened into a feeling—and then into a knowing. Then in a matter of a few minutes, I knew and felt with absolute certainty throughout all of my senses, that there was absolutely no reason that Freddie would not be absolutely and completely healed. In that moment of total certainty, Freddie got up and walked around as if nothing had ever happened—never again to return to his paralyzed condition.

Then one day a couple of weeks later, he came home from his daily territorial adventures with one eyelid shut. It remained that way for about two weeks. When it finally opened, I saw that what had been an eyeball now more closely resembled a piece of dead flesh; it had deteriorated probably to the point of no return. I again took Freddie to the vet, who gave him virtually no chance of retaining his eyeball. He gave me some ointment and wished me luck. He also advised me of the importance of returning in a few days to have the eyeball surgically removed. I left wondering if he would even survive such a radical procedure. I also worried that he might not survive it anyway; what if it was cancer?

Shortly thereafter, and again, while sitting on my deck with him, the spontaneous vision of Freddie's perfect health returned to me again. It quickly intensified to the point to where I saw it, felt it, and knew it with absolute certainty and clarity throughout all of my senses. His eye immediately began to heal, and was quickly and completely back to normal. It remained that way for the rest of his life.

I must emphasize that this was all completely spontaneous. Somehow it just happened; and it all seemed completely normal and natural at the time. I will also say that even though I was extremely grateful for Freddie's instant and seemingly miraculous healings, I had absolutely no idea of what had happened—or why it happened. I didn't think it or plan it or realize that any of it was even possible.

At the time I had absolutely no idea that the keys to my own healing—and yours too—were contained within this experience.

It took a long, long time before any awareness, any insights of what had happened began to come to me. In fact, in November 1988, Freddie quite suddenly went from good health to very ill; within a matter of a few days, he was dying and I was helpless to do anything about it. He soon passed as I sat there helplessly watching. It was a sad moment indeed; I would have given most anything for another spontaneous healing.

Finally, eleven years after these incidents—by 1991, I had internalized enough to begin to get the message; and I began to heal. What did I learn?

Looking back on it, clearly Freddie had no limiting beliefs about what was possible. He was at total peace and in harmony with his reality, no matter what his condition was; he had no resistance—absolutely none. Animals are not at war with themselves, only humans do that. He was simply a clear channel for the healing to pass through him. After all he was a cat; he was just being himself!

It was out of this remembrance of Freddie and his complete acceptance of the present moment—no matter what his circumstances were, that I finally began to realize that I was living a massive lie; and it's one that you may believe yourself. I believed that if I would just battle and try to change my condition enough, that somehow it would just magically clear up. Yet all of my attempts to alter or change my condition for the better only resulted in more setbacks and increased levels of frustration. I habitually made stress and painful low back feelings wrong, and tried to suppress them. You would too; we have all been conditioned to react as such. I felt quite helpless.

Finally, my inner light bulb turned itself on, and I began to realize that instead of being at war with my back, I had to learn to make peace with it, no matter what my condition was. Making peace with your life is not a function of putting up with or trying to change an unwanted condition. Rather, it is about learning to harmonize with it by learning how to flow with the natural currents of life. It is about aligning yourself with Universal Principles.

You've got to stop arguing with "what is" in your life, and here's why; you might think that it's your negativity that motivates you to be a good person or to make positive changes. You might think for example, that if you just make yourself wrong enough for being overweight, that that's going to motivate you to lose weight.

In fact, if that really worked well, then there wouldn't be many, if any overweight people. What actually happens is that when people make themselves wrong for being overweight, or for having any other problem, they have an uncomfortable feeling that primarily motivates them to get relief from the discomfort. It is a common practice for overweight people to eat in order to suppress the uncomfortable feelings they have about being overweight.

So, in order to eliminate the source of your distress so you can be more creative and at peace with your life, you must find a better way of relating to the reality of your experience. Otherwise, you're only going to get more of what you don't want—more suffering!

From this understanding, I could begin to put the pieces together into a composite whole. I was able to go back and reverse engineer my earlier experiences with Freddie, and develop a system that became the blueprint for my own healing. I have since shared it successfully with thousands of others; this was the birth of my Breath Alchemy Technique!

I must also say that to this day I still do not fully understand how Freddie was so spontaneously and completely healed; maybe it was pure luck, maybe my burning desire for his healing—and for my own healing was a factor. Whatever, he was a great teacher; he led purely by example, and with total humility.

I strongly suspect however, that it was out of my intention to see Freddie healed, and by being open to the possibility, that my Higher Self was able to come through me in a way that was not hindered by my limiting beliefs.

I have come to realize that most of us don't really know what I mean when I say that a burning desire in the form of one hundred per-cent intention was the entrance requirement for that level of certain knowing to come through me. So perhaps I can best illustrate this with the aid of the following true story.

Thomas Alva Edison who held 1,093 patents for diverse inventions, is responsible for great and brilliant creations such as the electric light-bulb, the phonograph, and the motion picture camera. He is known to be the most influential figure in the history of science, even though he failed hundreds of times.

His teachers considered him to be "too stupid to learn anything." He was labeled as "non-productive" by the companies he worked for.

Once a reporter, who'd come to see the demonstration of his newest invention, the electric bulb, asked: "How did it feel to fail a thousand times?"

To which he replied, "I didn't fail a thousand times. The light bulb was an invention with one thousand steps."

He failed over one thousand times before inventing the light bulb; still he was determined.

"I am not discouraged, because every wrong attempt discarded is another step forward."

He never gave up, he believed, "Nearly every man who develops an idea works at it up to the point where it looks impossible, and then gets discouraged. That's not the place to become discouraged."

That then, is what I mean by "burning desire!" If you are not burning with the fire of purpose, you will be stopped by your fears and your doubts when they come up—and they will come up! If you are on purpose, these same fears will seem unimportant and you will move through them.

In universal terms; what you currently believe is creating your life. If you're noticing that any of that is less than perfect for you, in order to change, your intention to change must be one hundred percent. Because the way it presently is, is the way you one hundred percent want it to be now. Nothing less than a hundred-percent intention to have it different, will be strong enough to overcome that. In universal terms, ninety-nine percent equals zero.

To translate that for a moment, it means for every intention, you also have a counter-intention; also known as a payoff. If your intention to heal is anything less than one hundred percent, the payoff will win every time. It has become a psychic entity and it wants to survive. Let's suppose for example that as a child, you discovered that when you were sick you received far more love and attention than you ever got for being well; by being sick, you got people to care, there's payoff number one! Payoff number two is you got to miss school; and there were times perhaps, where that came in quite handy.

All of that may have served you well at the time, but diminishing returns begin to set in later in life if you still have that subconscious program running.

If you can't feel it, taste it, be jumping up and down about it, it isn't enough. And the Law of Cause and Effect tells us that until you achieve the result, you know you don't want it. Having and wanting are simultaneous; the only way you know you really want it, is when you have it.

I was able to see, feel, and know Freddie's healing with total and complete certainty. And I recognized that my burning with the fire of purpose was the entrance requirement for that level of certain knowing to come through me. I was able to experience myself as my true nature; all consciousness, all awareness, all possibility, no form awareness, having a human experience. And that is you too!!!

Let me backtrack a bit here and emphasize that even though this level of certainty was well beyond believing that it was possible; it all begins with the belief that you can heal yourself. I'll come back to that, but first let's go back to the beginning and see if we can put all the pieces together. Otherwise, we are reduced to symptom masking and plastering over unresolved emotional trauma with positive thinking. Or you might just give up entirely and learn to live with it.

I wish to emphasize that there is a big difference between healing and finding a cure. If you have symptoms that are cured, that means you have been returned to the same state that existed before the symptoms surfaced. If you have not integrated the suppressed thoughts and feelings that caused the problem in the first place, then the symptoms will probably recur or perhaps recur in a different form.

Healing means that you have gone to the root-cause of the problem and eliminated it. The human body, like everything, is not physical. It's a holographic illusory physical form, which in its base state is waveform information. This has massive implications for true healing, because what the medical profession and Big Pharma do is treat the body; they only treat the symptoms. But the body is just a projection of its base state, which is waveform information. The causes are in the base state in information distortion. If you harmonize the distortion, the body *must* heal, because it's just a projection of its base state construct. Healing means you have gone to the source and harmonized the distortion in your body's base state; you have integrated the suppressed thoughts and feelings that have caused the symptoms.

All illness, disease, and even injury is the result of your negative beliefs and their accompanying unresolved emotions that are lodged in your subconscious mind. Anxiety, anger, grief, depression, and fear are all unresolved emotions; and your body reflects the state of your emotions. High blood pressure for example, is an outward reflection of irritation, resentment, hostility, and criticism. Arthritis is rooted in anger, criticism, resentment, and hostility. Longstanding resentment, grief, guilt, and self-condemnation can lead to cancer.

Your Higher Self, your connection to Source knows only Unity and the infinite perfection of the universe. As such it holds the idea of the perfect body in total complete and perfect health. However, if it flows through a consciousness of confusion, fear, resentment, futility, hostility, and irritation—then it will duplicate that state of affairs in your body. It will change to fit the tone of the directive coming from your subconscious mind. Your body is a holographic projection of your consciousness. It is a mirror image of your deep-seated subconscious beliefs; and that speaks to the importance of going down to the Lower Self and healing the inner child. If you would like to revisit that discussion, I will refer you back to chapters 19 and 20.

Pain is held in place by fear. If you allow yourself to relax into the fear sufficiently, you can discover that pain is held in place by specific, suppressed fearful thoughts. Two of these are fear of worsening and fear of irreparable damage.

With fear of worsening, you experience a body sensation—you make the sensation wrong, and become afraid that it will get worse. Then you tense or tighten the body as protection against the anticipated worsening pain. The pain in fact does get worse as a result of the increased tension and your fear is validated. With fear of irreparable damage, you are afraid that the pain will cause or has already caused damage that is beyond healing.

Chronic pain, disease, illness, or the old injuries that you have in your body are not actually in your body, they are in your mind. More specifically, they are a function of your perception, the unwitting context in which you are holding them. Pain for example means that there is an intensely felt sensation that you are making wrong. Until you change that, you will just keep recreating it, and the intensity level will increase. You can also surrender to a painful sensation and experience it as pleasurable. To do so will enable you to integrate the fear that was holding the "pain" in place and facilitate faster healing.

We know through science that every cell in the body is periodically renewed. Therefore, it is your consciousness, your state of awareness reflected through your subconscious beliefs that alters the cells. If you can see yourself not as your body, rather as pure consciousness, and that who you really are is infinite creative awareness that is manifesting reality, then you can start to take complete control over your body, your health, and your life.

Now let's go back to the idea that knowing with certainty must begin by changing your beliefs of what's possible—it begins with the belief that you can heal yourself. Right now, you probably have a deep-rooted belief that you can't, and the Law of Cause and Effect tells us that what you believe about your life is what controls your life, and that you always without exception get what you want (remember wanting is defined as both conscious and subconscious).

In our separate state where we don't connect the dots and see wholeness, we have become conditioned into believing that the power and answers to our problems lies outside of us. So we habitually give away our innate ability to heal ourselves to outside authority, namely doctors.

So even in the face of no evidence you begin by adopting the belief that it's possible to heal yourself. Belief when it's nourished widens into faith, then it becomes your experienced knowing.

Knowing with certainty is the next step; and that comes from changing your identity. You are *not* a human being having a spiritual experience; you *are* a spiritual being having a human experience! You must make this shift even in the face of no evidence for it, and you must be willing to come from there. In so doing, you become the space in which healing can occur.

You have created a context that turns what had been your unwitting, unconscious contexts coming from your reactive mind, into mere content, which can now be transmuted into the light of your conscious awareness. You must realize that you are always holding your life experiences in a context, so why not do it consciously.

You are infinite consciousness experiencing itself at a single point. Your Higher Self is your direct connection to this Source energy flowing through you. Your body is merely the necessary container to experience your true nature in this particular wavelength universe. So when you align yourself with your truest self, your body will begin to respond accordingly.

To enjoy perfect health, you must choose it and accept it. You must tune into the Source energy that's within you and let its healing power flow through your entire body. Prana *is* Source energy; it is the life-force energy in your body. Prana is the Sanskrit term used in Indian Yoga philosophy. It is also known as chi, ki, and even tachyon energy.

As I indicated in chapter 12, prana is absolutely essential for sustaining life as we know it. Prana is not the same thing as oxygen. Oxygen is carried to the cells of the body by the red corpuscles. Prana travels through subtle channels in the body that are called *nadis*. The acupuncture meridians are the main vessels for that flow, but they branch out and every cell in your body is nourished by the subtle channels of prana.

Your breathing is the main regulator of the flow of life-force energy (prana) in your body. The breath is the vital link between Spirit and matter, between pure Source energy (prana) and the physical universe. In a Breath Alchemy session, you are very aware of this life-force energy flowing through you. You can readily feel its soothing healing currents of divine love and know that every function of your physical being is whole and complete. That plus holding the idea of perfect health in your consciousness is the key to your healing. See yourself in your true form as pure consciousness. See yourself as radiant health, filled with joy and peace and doing everything you want to do in your life. This is who you are, so claim it!!

I have seen people heal just about every possible illness imaginable, from terminal cancer to minor illnesses, with the power of the Breath Alchemy Technique. In addition to having completely healed my back, I have also healed my respiratory system; and considering that my family has a long-standing history of emphysema, I considered this to be a major accomplishment! As a child, I had hay fever that kept me in a constant state of misery in the spring and summer, and far too many severe colds that would keep me in bed for periods ranging from a few days to a few weeks. I was diagnosed by some quack masquerading as a doctor, as being allergic to just about everything. That is now all history.

I'll conclude this chapter by sharing with you some powerful examples of recent healings, both with clients of mine and myself.

I was asked recently to do an interview on a podcast, where for the first time, I went public with my knowledge and experience of healing. About two weeks prior to the interview, I began to notice what first appeared to be a toothache. Since I never get toothaches, I decided to just monitor it. It soon began to feel more like it was an inner ear infection; and it continued to grow in intensity. By the morning of my interview, I was in pretty rough shape. Since I was about to speak publicly on the topic of healing, I decided I'd better practice what I preach. So, I gave myself a Breath Alchemy session; in that session, I was given perfect access to the feeling in my inner ear. Instead of resisting it, I was overjoyed, I was thrilled to have such clear access to this intense sensation. Holding the feeling in this context of total acceptance allowed me to easily integrate it by applying the Five-Step Harmonizing Method. It was a powerful, profound, and complete healing; and to this day, it has not in any way resurfaced. It also perfectly served me in the interview with a memory so fresh in my experience that I was able to share it in a most detailed and meaningful way.

I recently had a client, Carmen Wilde, who as a busy entrepreneur, in addition to being on the edge of burnout, was experiencing a serious recurring problem with persistent headaches. She went for a series of ten Breath Alchemy sessions with me; and we are both happy to report that her headache problem was permanently resolved during her first session. She also said that she has learned to relax on a much deeper level than ever before, so her burnout problem is history also. Carmen had this to say about our work together:

> At the time I signed up for Bob's Breath Alchemy program, I was experiencing severe headaches literally every day. For a driven entrepreneur like me with a very busy schedule the impact on my life was huge. I knew I had to do something very different to what I'd tried in the past and I've tried many alternative healing methods. I met Bob in Los Angeles in January 2018 and when I saw his passion for his work and heard about the results he gets, I knew instinctively that he could help me.
>
> I did ten weekly sessions with Bob and right at the outset his method had a positive impact. Bob is hugely passionate about his method, very patient in guiding you through the sessions and provides comprehensive information to help you understand the mechanics behind his method.
>
> What I love about the method is the extensive utility value, it's a powerful tool I use to increase energy levels, get into a focused state, relax, sleep better, stay composed, etc.
>
> And my headaches have literally disappeared! Wow! On the odd occasion when I feel a headache coming on (which by the way is only a fraction of the discomfort I used to experience), I go straight into circular breathing and it dissipates. If you're experiencing any kind of pain, anywhere in your body, Bob's Breath Alchemy can resolve it. I have a power technique for life and I'm deeply grateful to be enriched with such a life-changing method. Thank you Bob!

Then there is the case of Wilbur Albrecht, the guy who assisted me with the previous chapter. Wilbur, a doctor, was a burnout victim who was forced into retirement in his late forties after almost losing his eyesight. The following is his accounting of the results he achieved during his first year of working with me:

- Stopped drinking alcohol after years of frequent use.
- Stopped taking prescription sleep medication for bouts of insomnia.
- Stopped taking prescription medication for stress.
- Stopped daily morning coffee intake.
- Restarted Transcendental Meditation practice after years of not being able to do it because of "side effects."
- Shortness of breath and asthma almost nonexistent; stopped requiring frequent use of inhaler medications. This past summer was reported as being very bad for asthma too. I did not take steroid medication in 2015 that my doctor said I could benefit from.
- Sinusitis almost nonexistent; only one sinus infection with a fever in 2016 (had at least four cases in 2015 taking antibiotics after all four had at least two weeks of fever).
- Foot pain resolved. Saw a surgeon in 2015 for bilateral foot pain that came after daily walks and would frequently prevent me from sleeping. No medical or surgical treatment was received.
- Lost thirteen pounds (after having lost another thirteen pounds the year before). I was diagnosed with weight related pre-diabetes a few years ago indicating need for weight loss on my part to prevent diabetic illness.
- Blood pressure normal—avoided medication after a year of very high BP.
- Sky high triglycerides (in 2015); normal in 2016 taking only prescription fish oil.
- Liver function test normalized (abnormal in 2015)

- My retina specialist said there is some improvement in eyesight in my most damaged eye.

Here are more things I have noticed over the past year that I have added to my list:

- Notable increase in sense of well-being. People also notice I am happier and look healthier.
- Feeling more optimistic
- Feeling more supported by the universe; noting that beyond feeling this too, that things appear in my life from those around me that feels supportive.
- Feeling more sense of purpose; a greater feeling of connection to a "higher self" within me . . . a deeper part of my being where my true self resides.
- A better sense of knowing what I need to do now in my life (e.g., a more predominantly vegetarian diet, daily exercise, and regular yoga classes in the last year). Also more clearly acknowledging the benefit of a daily disciplined meditation practice in my life. Increased use of Reiki in my life (joined the regional Reiki association also).
- Realizing with greater clarity and confidence that I need to address my personal health issues with intention (and personal responsibility) and not just succumbing to a medical-system oriented approach.
- I feel five years younger; when I go to bed at night and awake in the morning in particular, I note, "Wow, I feel so much better physically and emotionally than I did a year ago." However immediately after a Breath Alchemy session I feel even younger than that (ten years younger easily). I have much more energy for daily exercise than I did before starting regular Breath Alchemy sessions.

- I note more emotional freedom; particularly in regard to relationships with those close to me and I notice more clarity also.
- In a group that I work with regularly I have been referred to as a leader; beyond just feeling confident I feel grounded in a deeper core part of my being–a very spiritual part of myself, especially right after a Breath Alchemy session that same day.
- Intuition is more notable than a year ago.
- Increased synchronicity attracting things in my life that help me to hear what I need to hear or to know things that would be helpful.

WITH MUCH GRATITUDE,
WILBUR ALBRECHT

26
We Are Nature!

"Those trees are your lungs. The earth recycles as your body. The rivers recycle as your circulation. The air is your breath. So what do we call the environment?"

DEEPAK CHOPRA

We are Nature, and we are Nature in the most profound way. We can't really draw a line and stand apart from it and say "That's Nature and this is us." We can't do that because we are Nature. Let me show you.

Let's begin with air and let's begin by following one breath. Now one percent of the air is an atom called Argon. Argon is an inert substance so it doesn't react with anything, so you breathe it in it goes into your body, and you breathe it out so it comes right back out again. So, it's a good atom to follow, and in case you're wondering how many Argon atoms there are in one breath. Well, there are three followed by eighteen zeroes; that's a lot of Argon!

So very quickly, everyone in your room is breathing zillions of Argon atoms in that one breath. And then suppose the door opens and out go billions of Argon atoms that everyone breathed. It goes around the Earth and one year later, if we were to come back into that same room, every breath would have about fifteen Argon atoms from that one breath you took a year ago. So that means then that every breath we take was once in the body of Jesus, every breath we take was once in the body of the Buddha, and if you go back sixty-five million years, every breath was once in the bodies of the dinosaurs. So air is this wonderful substance; it gives us life, and it connects us to all living creatures, past and into the future.

And water? Well, every one of us is at least sixty percent water by weight, so we're basically just a big blob of water. And yes, we have enough thickener added so we don't just drivel away on the floor. And as you know, our bodies leak water; we're losing it all the time. Yet the amazing thing is, our bodies know that. We don't have a big gauge, we don't need one, telling us that we're nearing empty, because our body knows exactly how much water is in it, and it's constantly keeping that level just right.

Water covers seventy percent of the planet and it evaporates; it forms clouds and it rains over the land, and it runs into rivers and lakes. And then it evaporates, and around and around it goes. So it cartwheels all around the planet, connecting us all together just as the air does, because we are water. And that water has come from all over the planet. So whatever we do to the water, we're doing it directly to ourselves.

How about Earth? The soil is what keeps us alive. We talk about how thin the atmosphere is, but the soil is just a very thin layer, and we are the Earth; we are the soil in the most profound way because most of the food we eat is grown in the soil. So you put the food in your mouth, you take the molecules out of the food, and you make it into what we are. So we are the Earth because we're created from molecules that plants have absorbed from the soil and we make it into our own body.

We are fire because every bit of the energy that makes us live and grow and reproduce; all of that energy in our bodies is originated as sunlight. Plants as we know, have found a way to capture this energy in the sunlight through photosynthesis. They are converting sunlight into chemical energy, which is sugar, and it can be stored. So that sugar holds the energy; and when we burn sugar, we release that energy. So we eat plants and we get that sunlight captured in the plants, and it makes up our bodies.

So, we are fire in the most profound way. And every bit of fire that we make, that is, outside of our bodies, every single bit of it was sunlight that was captured by plants and stored as fossil fuel, or stored as trees that we burn. Every bit of the energy that we use was sunlight. We are fire!

The inspiration for the above came from a lecture given somewhere by David Suzuki. I happened to catch it on some obscure cable channel. I was prepared though; I always keep a pen and notepad by my side.

Now since we are inseparably connected to Nature, let's look at some ways in which we can harmonize with the four elements.

Air

The breath is the vital link between spirit and matter, between pure energy and the physical universe. There is a life-energy that we are flowing in, that's flowing into us and through us, and that sustains us. Breathing is like our umbilical cord to our true mother, which is universal life. It is the umbilical cord to life itself; it's the bridge. You enter the physical world on an inhale, and you leave it on an exhale. Breath is that vital link between spirit and the body, it is a way of bringing Heaven onto Earth. It's the link between the spiritual and the material.

With Circular Breathing, the inner and outer breath merge. The outer breath being the lungs and physical breathing. The inner breath is the flow of life-energy or prana. In a Breath Alchemy session, you want those to join, you want your outer breathing to join with an internal rhythm, the inner rhythm of life. So you start out by noticing how you breathe naturally, and then expand it from there.

Everyone knows how to breathe, but to various degrees, their breathing has become inhibited. In general, peoples' breathing started to become inhibited at birth. At birth, you are trying to breathe, but the amniotic fluid is inside your system, so there is a certain amount of a feeling of choking. There are two things going on; one is that you have amniotic fluid in your mouth, nose, lungs, and throat. The other is that they want you to breathe, so they slap you on the butt to get you to start breathing. So you breathe in, but what there is, is liquid there. So it's like choking because you are trying to inhale, but at the same time you have to spit out the fluid, so you get a choking experience. What happens then is that our first experience of taking a breath is often paired with fear and pain.

It's like Pavlov's dogs where every day when it was time to feed the dogs, they would ring a bell, and then feed them. Pretty soon all they had to do was ring the bell, and the dogs would start salivating in the expectation of the food coming. Most humans have breathing and pain hooked up in a similar manner; and it is so subconscious that we don't even notice it. So, there is this deeply rooted, subconscious thought that breathing causes pain. That thought, as it lives in our bodies, works to stop the aliveness that we want more of.

Because it is beneath the surface, you may not be aware of any thoughts you might have that connect breathing with pain. But if you think about it, what's the first thing you do if you're walking along and you accidently stub your toe, and it hurts? You hold your breath. There is something in your body that says if I hold my breath, the pain won't increase. It doesn't go through your mind that way, but you automatically respond that way; just like Pavlov's dogs salivating at the sound of the bell.

Another example is if someone runs a stop sign while you're driving through an intersection, and you have to suddenly slam on your brakes in order to narrowly avoid a collision; what's the first thing you do before you put on the brake? You hold your breath. Also, if you are watching a sad movie, and you start feeling very sad, what do you do to make yourself not cry? You hold your breath and swallow.

So, people are constantly cutting off that supply of life and love without even realizing it. Breath Alchemy is the process of opening up your breath, and letting go of the association of breathing to pain to repair the experience of breathing, so you can experience more and more pleasure.

Most people then, are sub-ventilating, breathing less than they should be. Sub-ventilating anesthetizes the body. Emotions are energy in motion. Every time there's an energy that wants to be in motion, and you stop breathing, that energy has nowhere to go, so it stays in your body.

Eventually that stuck energy will start popping out in undesirable ways like aches and pains, stress and tension, stiff joints and arthritis, cancer, heart murmurs, and all the rest of it.

Breathing is one of the most powerful healing tools we have, and one of the least used. The body is like a pressure cooker; if the energy isn't allowed to move, it builds up and eventually pops out in undesirable ways.

Each one of the cells in our bodies holds our entire memory. The breath is like a flush, like what Drano does to a clogged drain. In a session, there is a certain amount of life-flow that is going through you, a life-stream. Ordinarily when we breathe, we barely touch that potential. Breath Alchemy is a way to increase the life-force in your body; there is a very distinguishable life-force energy, or love, that's doing the healing.

When you're breathing efficiently, you are breathing life-force energy (prana) as well as oxygen. This combination keeps the body clean—it flushes the nervous, circulatory, and respiratory systems as well as the aura or energy body. It cleanses physic dirt, negative mental mass, physical tension, physical illness, and emotional problems out of human consciousness.

The breath is a tool to transform your body. Instead of thinking that your body is a limitation that you have to transcend, you bring your body with you; you expand the energy in your body. It is called illumination, or transfiguration. The density of your body begins to dissipate because you're not holding. So you lighten up, also called enlightenment.

Any physicist will tell you that all we are is light, but the light gets so jammed up that it seems massive. If your cells are all jammed together, your breath is a way of separating them.

Water

Just how soothing is it to sit by a stream or creek and soak in all the negative ions? And how about swimming in the lake, or better yet, in the ocean; is that refreshing or what? And your shower or bath—here's a simple question for you, "Did you feel better before your bath or shower, or did you feel better after?" Stupid question, right?

We all know the value of daily bathing; yes it cleans the dirt off you, and it certainly leaves you feeling refreshed. But why is that? The answer is that it is also cleansing you of "psychic dirt"; as you relax, you can feel the stress and tensions from the day just melting away.

I recommend then, that you try bathing twice a day to clean the aura or energy body as well as the physical body. This can be done either by showers or by total immersion in a tub, the latter being a more efficient method of cleaning the energy body.

Water purification as an initiation into the spiritual life is an ancient practice; it is called baptism and it was in use long before the Christians began practicing it. Now we have the convenience of indoor plumbing and hot water heaters.

Let me remind you of the importance of filtering your bath or shower water. It's quite harmful to the body on an accumulated basis, to be bathing in chlorinated water. And the same is true for your drinking water.

Earth

Since we are Nature, it certainly makes sense to develop a conscious, harmonious relationship with Mother Earth. Standing apart, trying to conquer and control the natural world doesn't work; when we destroy Nature, we destroy ourselves. The Earth is alive. In order for you to be fully alive, you must reconnect with it, tune to it, and feel it. At a certain point, she will begin to communicate back to you.

There is a huge price to pay for living in cities with all the pavement, asphalt, traffic, pollution, and electromagnetic fields. In order to cope with it, you tune out, and in the course of doing so, you get numbed out. You lose your intuitive feel. Reconnecting with Nature is the essential aspect in getting that back.

So I suggest you spend as much time in Nature as you can; and while there, let walking or hiking become a regular part of your routine.

Fire

In the biblical book of Genesis, God promises Noah that the Earth will never be destroyed by flood again. He makes a covenant with Noah to seal this promise, saying that the rainbow will be the symbol of this covenant. The Bible says the next time the Earth is destroyed it will be by fire. In the New Testament book of II Peter there is a verse that says: "But the day of the Lord will come as a thief in the night; in which the heavens shall pass away with a great noise, and the elements shall melt with fervent heat, the earth also and the works that are therein shall be burned up."

If you believe the events of 1972, then you know we would have been destroyed by fire had not the Sirians intervened. Dare I say that were not evolved enough to just tune to the energy of the sun.

Fire is the most neglected of the elements. We are dangerously out of touch with it, and when we are out of touch with an element it can destroy us. Less than a century ago, fire was necessary for cooking and to provide warmth. Now we have central heating and microwaves. This is the same technology that is choking us with the internal combustion engine, and it is the same technology that has given us guns and weapons of mass destruction. All are examples of the misuse of the fire principle.

Sitting with a fire aids you in developing a conscious relationship to fire so you notice experientially what the fire is doing for you. The word "chakra" means wheel. Your wheels of energy are always turning. When you are by a fire, they are turning through the flames. Fire is the most efficient element for burning away anger and the death urge, the basic principle being anything that is not God gets burned away.

I would suggest that you go camping to remind yourself of just how pleasurable it is to sit in front of a good campfire. If you have a fireplace or a wood burning stove, use it. And if you don't, burn candles. We know that a room full of burning candles creates a wonderful atmosphere, a beautiful "Cathedral" type effect. So, give yourself the direct experience of fire.

27
Life Context

Once upon a time there was an entity named Vector Three. Vector Three is a composite entity of group souls on the eighth dimension. It is composed of three billion individual Spirits who are actively participating in the awaking of planets in 383 different worlds in five local universes.

How's that for an expansive idea? Well let's consider the possibility that it's more than just an idea.

Let's also consider the possibility that Earth is one of those 383 different planets, and that there are many millions of ETs here who are part of Vector Three. Since you are reading this, you just might be one of them.

Your job has been to come to this planet, become just like the indigenous life forms—in this case human—and then to go to sleep to forget who you are and forget why you are here, so you could become "just folks."

Because you are part of a planetary transition team, you needed to come here and go to sleep and then wake up. That is your expertise, and you have done the first part perfectly.

It is now time to remember that you are an interdimensional master, here to assist the planet in its awakening. You are here to co-create the birth of a new humanity!

The most important reminder as you awaken to your true self, your Higher Self, is to live your sacred purpose, and to follow your Spirit without hesitation; that is a prime directive as a transition team member.

Remember that you came from a world where everything is light and everyone feels connected to all of life. Everyone follows their Spirit without hesitation, there is only kindness, gentleness, and love there. You came with many friends to help Mother Earth in her ascension into the higher worlds.

Imagine for a moment; stretched across the line between dimensions are millions of Vector Three planetary transition team members—all like little fireflies, glimmering—right on the edge. All of them have agreed to come here to help the planet in its transition. They are like midwives in a birth. And you're all ready to go. And you're looking around at your best buddies and you say "Well once more into the breach, here we go. We've done this on many different planets."

And you look around at your best buddies, knowing for a while, you're going to lose contact with them. You're going into the darkness; you're going into unconsciousness on purpose. And deep inside you, you know that once again you'll find each other.

And you will bring forth something that has not been here in a very long time—a new age so that humanity can return to the higher levels of consciousness it had once achieved.

And just before you shoot off inter-dimensionally and show up here in babies' bodies or walk into an adult body, you give each other the old ET thumbs up and you say "See you on planet Earth!" And you shoot off and you come down here.

If you are willing to consider the possibility that you are a master, and that you are here to bring your love, light, and wisdom here, there are some useful elements to key into the universe to help that happen. The entire universe rearranges itself to accommodate your picture of reality. Whatever picture of reality you give, the universe must manifest it. No two people have the same picture of reality, and yet the universe rearranges itself to accommodate each and every one of those realities.

It is also very useful to burn in your heart with the fire of purpose. If you are, you will do whatever it takes to co-create Heaven on Earth and to awaken to the master that you are.

One of the major elements of your picture of reality is your identity. It takes an enormous amount of veiling to create the illusion that you are just a powerless individual. As such, you have likely been turning to outside authority for advice, and completely negating the fact that you are a spiritual being having a human experience! You have to start with no evidence that this is real. At some point you just have to declare that you are a master. The entire universe will then begin to rearrange itself to accommodate you.

In addition to your identity, there are two other main elements to your picture of reality, one is the context in which you live your existence, and the other one is the way in which you measure reality.

If you identify yourself as a master, then you will treat your experiences as a remembering, rather than a learning process. Masters know that they are all possibility; humans are here to see if they can learn enough to become enlightened.

So, begin to proclaim that you are a master, begin to shift your life context by shifting the questions that you ask of your experiences. Ask how does this contribute to the co-creation of Heaven on Earth, instead of what am I supposed to learn from this?

Then shift the way in which you measure reality. From the third dimension, the way of measuring reality is beliefs and experiences. As you progress you measure it from your feelings, your intuition, and your abilities to read subtle energies. From the higher overtones of the fourth dimension, which is where we are headed, things are real because you say so. Whatever you say is reality, the universe will instantly manifest.

What if you suddenly became totally fascinated with your picture of Heaven on Earth? If you took a radical shift in your identity, in how you contextualize your reality, and in how you measure it, the universe would have no choice but to give you a radically different reality.

After you have done all of that, then all you have to do is follow your Spirit without hesitation. You are living your sacred purpose! Your life will then flow as you fiercely hold your vision. You will, as Gandhi said, "Be the change you wish to see in the world."

Okay I hope this is starting to take root, because I'm running out of ways to talk about it. Our true nature is that we are interdimensional masters—we exist on different dimensions simultaneously. In our higher aspects, then, we already are Christ-consciousness beings. We are spiritual beings having a human experience here in the third dimension. Having that awareness creates the possibility of living your life in its truest context.

Our mission is to wake up to our true nature so we can get on with it, so we can do what we came here to do. Our true purpose is to bring our higher-dimensional light and wisdom here to co-create Heaven on Earth, to assist the birth of this planet into the fourth dimension.

And the process is just like birth; in fact, it is birth. The more awake and conscious we are, the more we will be able to gently and wisely assist the planet. It can be a safe and exciting journey. Contrast that with the alternative. If we are asleep and filled with fear, then the birth will be a reflection of that—an experience filled with fear, pain, and struggle.

The entire planet, and everyone on it, is in the midst of this "birth." We are all mutating—Mother Earth is in transition and so are we individually. Some of us are more aware of this than others. It means we get to experience the joy of having more light flow through us and also the inconvenience of having our old world dissipate before the new one is completely together.

Like it or not, we are all mutating, and if we resist, it can get pretty rough. If you try to go back to your old way of being, it won't work anymore. If you begin to identify as a master here to co-create Heaven on Earth, you are moving in the direction of the transformation and therefore helping it. You will be able to find your way more easily.

We do create our reality unerringly. In fact, all we ever see "out there" is simply an outward reflection of our own inner state of affairs. If we see fear and limitation, it is because that is what is going on within us.

As we begin to wake up to our true nature and identify with the master that we are, this then becomes our inner experience. We are then able to project it outside ourselves; and as we do this, we are indeed co-creating Heaven on Earth. We do have the ability to change the future. It is all a function of our consciousness.

The second coming of Christ is not the second coming of a man; rather, it is an emanation of the luminescence within all beings who are ready. The beings on this planet are becoming illuminated from within, and they are learning to rely on their own internal authority.

As this happens, the planet has less work to do. At a certain point, an exponential level will be reached, the leap in dimensionality and con-sciousness will take place, and the planet will become lit from within.

We are not going to ascend and leave; we are going to ascend into light bodies and stay, but on the fourth dimensional level. *Bon Voyage!*

28
What Does It All Mean?

As mentioned previously, the new grid, the one that is allowing us to move into Christ-consciousness, was completed on February 4, 1989. You could think of it as an electromagnetic, geometrically shaped fishnet that extends around the entire globe. Its geometrical nature is based on an icosahedron and a dodecahedron. If you look at an icosahedron, you will see that it is composed of pentagonal faces that come out to a point. You get to that "point" by putting five tetrahedrons together, which makes an "icosahedral cap." If you put an icosahedral cap on each of the twelve pentagonal faces of a dodecahedron, you will have the exact configuration of the grid. Since the icosahedron is associated with water and the dodecahedron with prana, you could also think of the grid as a combination of water and prana.

This grid used to be around our planet but was destroyed when the failed external Merkaba of the Martians ripped open the dimensional levels of the Earth. This happened sixteen thousand years ago, prior to the sinking of Atlantis. Because we had attained this level previously, we were given permission to synthetically recreate it.

When we fell, we stopped breathing through the tube, causing the prana to bypass our pineal gland, thus creating a polarized world in which we experience judgment and separation. We moved from our heart, which knows only unity, into our mind, which knows only polarity.

Since the dawning of the new millennium, two events have occurred that are changing everything. The first is the movement of the Earth's Kundalini, which happens every thirteen thousand years and is tied to the precession of the equinoxes.

The Kundalini, which is related to the Earth's spiritual growth process, has two poles. One is in the center of the Earth, while the other is somewhere on the surface. Every thirteen thousand years its polarity changes to the opposite pole. As the Kundalini energy comes out of the center of the Earth it moves up in a wave, much like the motion of a snake. When it reaches the surface, it moves all over the Earth until it locates a new spot.

Once located in Lemuria, the surface pole relocated to the southwest corner of Atlantis until it moved into the western mountains of Tibet thirteen thousand years ago. It began its latest journey in 1949 and finally settled in the northern mountains of Chile in 2002.

Every time this pole moves it has a new vibration, which in turn takes us to a new level of consciousness. According to Drunvalo:

To the few who know of this event and what is occurring all around us, a wisdom is transferred, and a peaceful state of being becomes their inheritance, for they know the awesome truth. In the midst of chaos, war, starvation, plagues, environmental crises, and moral breakdown that we are all experiencing here on Earth today at the end of this cycle, they understand the transition and know no fear. This fearless state is the secret key to the transformation that for millions of years has always followed this sacred cosmic event.

On one level, this means that spiritually the female will now have her turn to lead mankind (womankind) into the New Light. And eventually, this female spiritual light will permeate the entire range of human experience, from female leaders in business and religion to female heads of state.[1]

When this "Serpent of Light" moved into Chile in 2002, it was not working. Directly above it, the new grid was distorted, and this kept the Kundalini from functioning. Corrections to the new grid were made for the last twenty years or so, and once this final correction was completed, the Kundalini began to function.

The second major event of the new millennium was the birthing of the new grid in January 2008. Its conception began 13,200 years ago when Thoth, Ra, and Araragat began the synthetic reconstruction of the grid in an area of the Giza plateau. The recent birth of the completed synthetic grid lasted about one month, with the assistance of Polynesian elders.

The Christ-consciousness grid has been fully formed and birthed. It is now alive and conscious; it is a living energy field around the Earth, and this changes everything. In every instance, whenever a synthetic grid has become a living entity and connected to its planet, it has always gone to the next level. Mother Earth has made a conscious decision to move into the higher overtones of the fourth dimension.

These two events have put us into a different potential world. We are now in a heart-based energy field; when you connect to it, a new possibility opens up. This means that our spiritual acceleration will quicken dramatically—there is nothing left to stop it.

It also means that sometime, perhaps soon, the entire cycle that we are now on will disappear in a single day. In its place a whole new world will be birthed, one based not on the mind, but rather on the heart. We are right on the edge of the emergence of this world. Everything is in place; the Kundalini has moved; the grid is alive. Yet most of the world continues in its old pattern, thinking this is the way it will be forever, hardly imagining that something incredible is about to occur.

This is a time of great celebration as we move out of the darkness and into the light. It means that the veils will be lifted, we will remember and live our intimate connection to all life, we will be allowed to reunite with our cosmic brothers, and to move about the

Universe. We will completely redefine what it means to be human!

It appears that we have made it in a really big way, and we will soon find ourselves not only in the higher overtones of the fourth dimension, which is the first step, but far beyond. We will continue to move up through the dimensional levels until at some point, soon, we will move through the great wall and into the next octave. This is a totally unheard-of event in the history of our universe.

There also appears to be a resolution in progress to the age-old question of Lucifer versus the light. This is a situation that for a long time appeared to be heading toward total chaos in exactly the same way it had in the previous three times. God planted a seed and created a situation—the 1972 Sirian experiment—where there was only one choice. It came from technology onto a Lucifer-rebellion planet (Earth). This experiment and its results now seem to be coming up with the answer for both sides.

For a long time, it was necessary for Michael and Lucifer to struggle against each other. It was necessary because it had a function beyond what you would normally think. The forces of "good" and "evil" are constantly approaching you from all different directions and have a greater purpose that has to do with the timing of events. God works though both factions to achieve a sense of timing. In other words, the dark force, the dark brotherhood, does everything it can to make sure that you don't evolve; it does everything it can to stop you. The light force does everything it can to get you to raise up and move higher in consciousness. On a Unity level, they are working together, so that you move at exactly the right moment in time.

Because we have gone into polarity or duality consciousness, we have broken this down into "what is good" and "what is bad" and we now judge everything. From a higher level of life, all that is occurring is God. There's really nothing else. There is an absolute Unity that has always been and will always be, and we can't see it because we chose this particular pathway, which was "right" at the time.

Now we have to learn to move in a different direction. It's already occurring on the higher levels of life; Lucifer and Michael have made an agreement, and the dark and the light forces are now merging into oneness again. The war or the struggle that had gone on for so long which was necessary has now become unnecessary. A Unity that we once knew a long time ago is going to begin to move through the Earth. It's already occurred on the higher dimensional levels of the Earth. It comes down through the planes and it eventually manifests down here.

What it means is that we must learn to drop all judgments—just eliminate them and come from a different place where we see that no matter what's occurring, there is absolute perfection in it, that God is in every moment no matter what is happening, and that there is a higher purpose in it. This is not easy because the consciousness of "good" and "evil" is engrained in us so greatly; it's in every cell in our body. The key to letting your judgments be, as I showed you in chapter 22, begins with the recognition that every make-wrong or judgment has a corresponding energetic component in your body—a feeling. And because the judgment carries a strong negative charge with it, the corresponding emotions (feelings) will be negatively charged and will feel unpleasant. And we have been well conditioned to do our utmost to avoid unpleasant feelings at all cost.

In recalling the similarity between the personal light-versus-dark battle inside you and the age-old struggle between the Great White Brotherhood and the Great Dark Brotherhood, even though they both appear to be in opposition, they are actually working together. In this context your internal "darkness" is trying to get your attention so you can revisit these feelings and integrate them into your sense of well-being. I will refer you back to chapter 22 for the details on how this is done. As you begin to do this, you will experience wholeness again, and you are going to find something very amazing will begin to happen. The higher Spirit of God will begin to move through you in a brand-new way in all that you do.

Thanks to the dance between Archangels Lucifer and Michael, a situation was created, taking what had always been a very slow process into super acceleration. It is now beyond all the players; something greater than the created hierarchy has stepped forth and is expressing itself.

The trigger could have been the all-encompassing attention we have received. God created a situation so interesting that all life just had to look and, as we know, the observer affects the outcome of the experiment.

There was a time before we created this universe when you and I were One. Once we created this external universe, we decided to step into it in order to experience it directly. Great Spirit split himself/herself into two. Part of him remained outside this created experiment, while another part of the Spirit of God—using the eternal pattern of creation, the Merkaba—moved in to live it.

It appears that God (us on the other side of the wave form universe) has intervened directly. We appear to be heading for a level of existence that is beyond our ability to even imagine, the universe is being recalled, we are going back home!

Notes

Chapter 1:
First Contact

1. George Knapp (narrator), *UFOs: The Best Evidence* (Las Vegas, NV: KLAS TV, 1989), video.
2. Associated Press, "Flying Disc Found; In Army Possession," *The Bakersfield Californian*, July 8, 1947, p. 1.
3. Milton William Cooper, *Behold a Pale Horse* (Flagstaff, AZ: Light Technology Publications, 1991), 208.
4. Cooper, *Pale Horse*, 208.
5. Milton William Cooper, *The Secret Government: The Origin, Identity, and Purpose of MJ-12*, manuscript (Huntington Beach, CA, 1989).

Chapter 2:
What's Going On?

1. Knapp, *UFOs*.
2. Knapp, *UFOs*.
3. Knapp, *UFOs*.
4. J. Randolph Winters (narrator), *The Pleiadian Connection* (The Pleiades Project, 1991), video.
5. BGR Entertainment Corp., *Contact* (Phoenix, AZ: Savadove Young Films, 1978), video.
6. BGR Entertainment Corp., *Contact*.

7. Bart Sibrel (writer/director), *A Funny Thing Happened on the Way to the Moon* (2001), video.

8. Sibrel, *Funny Thing.*

9. Zohar Star Gate TV, *Moon Landing Discoveries That Show Man Has NEVER Set Foot on the Moon* (2018), video.

Chapter 3:
Why Now?

1. John White, *Pole Shift* (Virginia Beach, VA: A.R.E. Press, 1980), 94.

2. White, *Pole Shift*, 149.

Chapter 4:
Problems with Planet Earth

1. Drunvalo Melchizedek, "Dry/Ice: Global Warming Revealed," Spirit of Maat website, July 19, 2004.

2. Robert Hunziker, "Fukushima Darkness," Democracy Press website, November 22, 2017.

3. Mike Adams, "Solar Flare Could Unleash Nuclear Holocaust across Earth, Forcing Hundreds of Nuclear Power Plants into Total Meltdowns," Natural News website, September 13, 2011.

4. Satya Nagendra Padala, "Solar Storms Could Disrupt Earth This Decade: NOAA," International Business Times website, August 8, 2011.

Chapter 6:
Earth History

1. David Icke, *The Lion Sleeps No More* (David Icke Books, 2010), DVD.

2. John Anthony West, "Civilization Rethought," *Conde Nast Traveler*, February 1993.

3. Robert M. Schoch, "Redating the Great Sphinx," *KMT, A modern Journal of Ancient Egypt* 3, no. 2 (1992): 52–59, 66–70.

4. Zecharia Sitchin, *Genesis Revisited* (New York: Avon Books, 1990), 15.

5. Sitchin, *Genesis Revisited*, 19.

6. Eberhard Schrader, from *Die Kielschriften und des alte Testament*, quoted by Zecharia Sitchin in *The 12th Planet* (New York: Avon Books, 1976), 209.

7. Joseph E. Mason, "'The Code' of Carl Munck and Ancient Gematrian Numbers Part One," Great Dreams website, 2008.

8. Drunvalo Melchizedek, *The Mayan Ouroboros* (San Francisco: Weiser Books, 2012), 119–20.

Chapter 9:
The Right Eye of Horus

1. Stan Tenen, *Geometric Metaphors of Life* (The MERU Foundation, 1990), video.

Chapter 13:
The Philadelphia Experiment

1. Preston B. Nichols and Peter Moon, *The Montauk Project: Experiments in Time* (Westbury, NY: Sky Books, 1992).

2. Al Bielek and Vladimir Terziski, "The Philadelphia Teleportation and Time-Travel Experiments of the Illuminati," American Academy of Dissident Sciences, May 1992, video.

3. Richard C. Hoagland, *The Monuments of Mars: A City on the Edge of Forever*, Berkeley, CA: North Atlantic Books, 1987.

4. Richard C. Hoagland, *Hoagland's Mars, Volume 1: The NASA-Cydonia Brieefings* (Hoagland and Curley, 1991), video.

5. George C. Andrews, *Extra-Terrestrial Friends and Foes* (Libum, GA: IllumiNet Press, 1993), 211.

6. Linda Moulton Howe, *An Alien Harvest: Further Evidence Linking Animal Mutilations and Human Abductions to Alien Life Forms* (Littleton, CO: Linda Moulton Howe Productions, 1989).

Chapter 14:
1972

1. Anthony R. Curtis, *The Space Almanac* (Woodshire, MD: Arcsoft, 1990), 607.

2. Kerby Ferrell. "An Active Sun in a Normally Quiet Period." *Science News* 102, no. 8 (August 1972): 119.

Chapter 15:
The Secret Government

1. Preston B. Nichols and Peter Moon, *The Montauk Project: Experiments in Time* (Westbury, NY: Sky Books, 1992), Appendix E.
2. Gordon-Michael Scallion, "UFOs From Earth," *Earth Changes Report* 2, no. 4 (May 1992).
3. Al Bielek, quoted from an interview by Susanne Konicov, *The Connecting Link,* no. 19.
* Further resources for this chapter were *Science Report,* "Alternative 3," from a broadcast of Anglia Television Limited, Norwich, England, April 1, 1977, Anglia Productions, written by David Ambrose, produced by John Rosenberg and Milton William Cooper, "The Secret Government, The Origin, Identity, and Purpose of MJ-12" (Huntington Beach, CA: Manuscript copyright 1989).

Chapter 16:
End-Time Prophecies

1. Frank Waters, *Book of the Hopi* (New York: Penguin Books, 1963), 3.
2. Scott Peterson, *Native American Prophecies: Examining the History, Wisdom and Startling Predictions of Visionary Native Americans* (New York: Paragon House, 1990), 159.
3. Peterson, *Native American Prophecies*, 160.
4. Peterson, *Native American Prophecies*, 160.
5. Peterson, *Native American Prophecies*, 161.
6. Waters, *Book of the Hopi*, 167.
7. Robert Ghost Wolf, *Last Cry: Native American Prophecies, Tales of the End Times* (Spokane, WA: Mistyc House Publishing, 1993), 73.
8. Ghost Wolf, *Last Cry*, 74–79.
9. "Comet Holmes Photogallery," SpaceWeather website.
10. "Interview with Drunvalo Melchizedek," Conscious Media Network website, May 2008.
11. "The Great Event Is Coming," Raphael's Healing Space website, March 12, 2018.

Chapter 18:
Our "Illusory" Holographic Universe

1. Luke Rhinehart, *The Book of est* (New York: Holt, Rinehart and Winston, 1976), 169.
2. Rhinehart, *Book of est*, 172.
3. Rhinehart, *Book of est*, 173.
4. Rhinehart, *Book of est*, 175.
5. Rhinehart, *Book of est*, 177.

Chapter 22:
Breath Alchemy

1. *Flower of Life 2000+ Workshop Student Manual* (Flower of Life Research, LLC, 1999), 34.

Chapter 28:
What Does It All Mean?

1. Drunvalo Melchizedek, *Serpent of Light: The Movement of the Earth's Kundalini and the Rise of the Female Light, 1949–2013* (San Francisco: Weiser Books, 2008), 10.

Bibliography

Adams, Mike. "Solar Flare Could Unleash Nuclear Holocaust across Earth, Forcing Hundreds of Nuclear Power Plants into Total Meltdowns." Natural News website, September 13, 2011.

Andrews, George C. *Extra-Terrestrial Friends and Foes*. Libum, GA: IllumiNet Press, 1993.

BGR Entertainment Corp. *Contact*. Phoenix, AZ: Savadove Young Films, 1978, video.

Bielek, Al, and Vladimir Terziski. *The Philadelphia Teleportation and Time-Travel Experiments of the Illuminati*. The American Academy of Dissident Sciences, May 1992, video.

Cooper, Milton William. *Behold a Pale Horse*. Sedona, AZ: Light Technology Publishing, 1991.

———. *The Secret Government: The Origin, Identity, and Purpose of MJ-12*. Self-published, 1989.

Curtis, Anthony R. *The Space Almanac*. Woodshire, MD: Arcsoft, 1990.

Erhard, Werner. *Parents: The Fundamental Relationship*. 1981, audio tape.

Ferrell, Kerby. "An Active Sun in a Normally Quiet Period." *Science News* 102, no. 8 (August 1972): 119.

Fetzer, James. "Proof That the Pandemic Was Planned & with Purpose." Principia Scientific International website, September 30, 2020.

Fuller, Buckminster. "A Candid Conversation with the Visionary Architect/Inventor/Philosopher R. Buckminster Fuller." *Playboy* 19, no. 2 (February 1972): 15.

Ghost Wolf, Robert. *Last Cry: Native American Prophecies, Tales of the End Times*. Spokane, WA: Mistyc House Publishing, 1993.

Hoagland, Richard C. *"Hoagland's Mars Volume 1: The NASA Cydonia Briefing.* Hoagland and Curley, 1991, video.

———. *The Monuments of Mars: A City on the Edge of Forever.* Berkeley, CA: North Atlantic Books, 1987.

Howe, Linda Moulton. *An Alien Harvest: Further Evidence Linking Animal Mutilations and Human Abductions to Alien Life Forms.* Littleton, CO: Linda Moulton Howe Productions, 1989.

Hunziker, Robert. "Fukushima Darkness." Democracy Press website, November 22, 2017.

Icke, David. *The Lion Sleeps No More.* David Icke Books, 2010, DVD.

Knapp, George (narrator). *UFOs: The Best Evidence.* Las Vegas, NV: KLAS TV, 1989, video.

Mason, Joseph E. "'The Code' of Carl Munck and Ancient Gematrian Numbers Part One." Great Dreams website, 2008.

Melchizedek, Drunvalo. "Dry/Ice: Global Warming Revealed." Spirit of Maat website, July 19, 2004.

———. *Serpent of Light: The Movement of the Earth's Kundalini and the Rise of the Female Light, 1949–2013.* San Francisco: Weiser Books, 2008.

———. *The Mayan Ouroboros.* San Francisco: Weiser Books, 2012.

Nichols, Preston B., and Peter Moon. *The Montauk Project, Experiments in Time.* Westbury, NV: Sky Books, 1992.

Padala, Satya Nagendra. "Solar Storms Could Disrupt Earth This Decade: NOAA." International Business Times website, August 8, 2011.

Peterson, Scott. *Native American Prophecies: Examining the History, Wisdom and Startling Predictions of Visionary Native Americans.* New York: Paragon House, 1990.

Rhinehart, Luke. *The Book of est.* New York: Holt, Rinehart and Winston, 1976.

"Roswell Incident." Wikipedia, November 2008.

Scallion, Gordon-Michael. *Earth Changes Report.* Westmoreland, NH: The Matrix Institute, 1992.

Schoch, R. M. "Redating the Great Sphinx." *KMT, A Modern Journal of Ancient Egypt* 3, *no. 2* (1992): 52–59, 66–70.

Sibrel, Bart. *A Funny Thing Happened on the Way to the Moon.* 2001, video.

Sitchin, Zecharia. *Genesis Revisited.* New York: Avon Books, 1990.

———. *The 12th Planet.* New York: Avon Books, 1976.

Temple, Robert. *The Sirius Mystery.* New York: St. Martin's Press, 1976.

Tenen, Stan. *Geometric Metaphors of Life*. The MERU Foundation, 1990, video.

Waters, Frank. *Book of the Hopi*. New York: Penguin Books, 1963.

West, John Anthony. "Civilization Rethought." *Conde Nast Traveler*, February 1993.

White, John. *Pole Shift*. Virginia Beach, VA: A.R.E. Press, 1980.

Winter's Randolph (narrator). *The Pleiadian Connection*. The Pleiades Project, 1991, video.

Zohar Star Gate TV. *Moon Landing Discoveries That Show Man Has NEVER Set Foot on the Moon*. 2018, video.

Index

About the Author

BOB FRISSELL is the founder of the Breath Alchemy Technique and has been a teacher for more than thirty-six years. His books are regarded as underground spiritual classics. In addition to *Nothing in This Book Is True, But It's Exactly How Things Are*, he is the author of *Something in This Book Is True* and *You Are a Spiritual Being Having a Human Experience*. His books are published in twenty-five languages and are available in more than thirty countries. Musicians have credited the ideas presented in Bob's books as a source of inspiration for their own creative work. This list includes Tool, Danny Carey, and Gojira.

Bob is on a mission to help as many thousands of men and women as he can, in opening them up to their unlimited potential by discovering that the resolution to any unwanted condition lies within. He gives private Breath Alchemy sessions on Skype and in person, along with coaching consultations on Skype.

Bob has been a featured speaker at the Global Congress of Spiritual Scientists in Bangalore, India, the 4th Annual Symbiosis Gathering at Yosemite, the Prophets Conference in Tulum, and many

New Living Expos. He has also appeared on numerous talk shows, including *The Jeff Rense Program*, *Red Ice Radio*, and *Far Out Radio*, and he has been a three-time guest on *Coast-to-Coast AM*. He has presented his workshops throughout North America, Australia, and Europe.

On a personal note, Bob is a Nature enthusiast. He loves to hike, and all manner of cats, squirrels, ducks, wild turkeys, deer, and redwood trees catch his eye. He lives in Sonoma, California.

For further information or to contact the author for a consultation, visit the author's website, BobFrissell.com.